UNIVERSITY SERIES IN MODERN ENGINEERING

Managing Editor:
A.V. Balakrishnan
Department of System Science
University of California
Los Angeles, California 90024
USA

SYSTEMS AND SIGNALS
N. Levan
1983, iv + 173 pp.
ISBN: 0-911575-25-1

LINEAR DYNAMIC SYSTEMS:
THE ELEMENTARY STATE SPACE THEORY
A.V. Balakrishnan
1983, approx. 200 pp.
ISBN: 0-911575-27-8

STOCHASTIC FILTERING AND CONTROL
A.V. Balakrishnan
1983, approx. 200 pp.
ISBN: 0-911575-26-X

ELEMENTS OF NON-LINEAR PROGRAMMING
N. Jacobsen
1983, approx. 220 pp.
ISBN: 0-911575-28-6

UNIVERSITY SERIES IN MODERN ENGINEERING

SYSTEMS

AND

SIGNALS

N. LEVAN

OPTIMIZATION SOFTWARE, INC.
PUBLICATIONS DIVISION, NEW YORK

Author
Nhan Levan
System Science Department
University of California
Los Angeles, California 90024
USA

Library of Congress Cataloging in Publication Data

Levan, N.
  Systems and signals.

  (University series in system science; 1)
  Bibliography: p.
  Includes index.
  1. Signal theory (Telecommunication) 2. System analysis.
I. Title. II. Series.
TK5102.5.L48 1983 621.38'043 83-4142 ISBN 0-911575-25-1

Worldwide Distribution Rights by Springer-Verlag New York, Inc., 175 Fifth
Avenue, New York, New York 10010, USA and Springer-Verlag Berlin-
Heidelberg-New York, Heidelberg Platz 3, Berlin-Wilmersdorf 33, West Ger-
many.

ISBN: 0-911575-25-1   Optimization Software, Inc.
ISBN: 0-387-91222-3   Springer-Verlag New York-Heidelberg-Berlin
ISBN: 3-540-91222-3   Springer-Verlag Berlin-Heidelberg-New York

About The Author:

Nhan Levan obtained his Ph.D. in Electrical Engineering from Monash University in Australia in 1966, and has been with the School of Engineering and Applied Science at UCLA since then. Currently he is Professor and Vice-Chairman in the Department of System Science. He was the recipient of the UCLA Distinguished Teaching Award in 1973, and of the American Society of Engineering Education Western Electric Fund Award for "Excellence in Engineering Education" in 1979.

PREFACE

This book has been designed to serve as a text for a Junior-level course in Engineering. It has been used in the Undergraduate Engineering program at UCLA for over six years and has gone through many revisions.

The prerequisites are two years of Calculus, including Differential Equations, and two years of Physics, including Electricity.

There are five chapters which can be covered at a reasonably comfortable pace in one quarter (10 weeks, or approximately 40 contact hours). Chapter 1 begins with the fundamental notion of a System and its Input-Output description, and proceeds quickly to the main properties of the Input-Output transformation. Chapter 2 features the Impulse Response function and its role in the time-domain analysis of linear systems. The Laplace Transform is introduced in Chapter 3, as a tool in the s-domain analysis of linear dynamic systems.

We turn next to Signals, beginning with Fourier Series analysis in Chapter 4. One noteworthy item here is that Mean Square approximation is introduced via the Orthogonality Principle. More advanced topics, such as Fourier Transforms and the Sampling Principle for Band-limited Signals are covered in Chapter 5.

An abundant supply of problems is provided. Each chapter concludes with a set of problems covering the chapter-material. In addition a hundred review problems are listed in the Appendix.

Nhan Levan
Los Angeles
March, 1983

CONTENTS

Dear Shane;

This is the book I promised to send you. Hope you will be able to use it. Love from —

Our love to everybody.

* hope this is the correct spelling

CHAPTER 1. SYSTEMS: THE INPUT-OUTPUT DESCRIPTION

This chapter is devoted to basic properties
of systems with an input-output description.
Inputs and outputs -- or signals -- are
always taken to be deterministic functions
of time.

SYSTEMS: MODEL AND MATHEMATICAL MODEL

The term system is often used loosely. Although a
precise definition would be cumbersome, for our purposes
we may regard a system as characterized by "inputs" ("sti-
muli") and "outputs" ("responses").

To study a system one generally begins by forming a
model for it. A model is an idealized version of the system and, of course, is not necessarily unique. In other
words, a system can admit more than one model, depending
on the uses envisaged. Here we are interested only in
models described in mathematical terms. Such a description
is often called a mathematical model of the system.

For our purposes, a system is represented by a closed
box with a number of accessible terminals as depicted in
Figure 1.1. Terminals are divided into two groups: input

Figure 1.1.

1

terminals and output terminals. At input terminals, inputs are applied to the system, while outputs are observed (or measured) at output terminals.

Inputs and outputs -- or signals -- are taken to be time functions, i.e., numerical functions of the time variable t, and are written as $x(\cdot)$ and $y(\cdot)$, respectively. Moreover, they are also deterministic, meaning their values $x(t)$ and $y(t)$ are completely specified for each value of the time variable t.

Let $x(\cdot)$ and $y(\cdot)$ be inputs and outputs of a system, then the pair of functions $(x(\cdot),y(\cdot))$ is called an "input-output" pair. If X and Y are allowable input and output families, then the system is completely characterized by its input-output data, i.e., the family {$(x(\cdot)$, $y(\cdot)$, $x(\cdot)$ in X and $y(\cdot)$ in Y}. This is the most general description of a system. Note that, in general, the "output" need not be uniquely determined by the "input"; the same input $x(\cdot)$ may have several outputs $y(\cdot)$ in the family {$x(\cdot),y(\cdot)$}.

A system is said to have an input-output description if its output can be expressed completely in terms of the corresponding input function, i.e., for each $x(\cdot)$ there is only one $y(\cdot)$. Such a description conveys the idea that the output is "caused" by the input: the system transforms each input into a unique corresponding output. Thus for a system with an input-output description we can represent its "action" by an (abstract) transformation $T[\cdot]$

2

acting on an input $x(\cdot)$ to give a unique output $y(\cdot)$:

$$y(\cdot) \;=\; T[x(\cdot)]\;, \qquad x(\cdot) \text{ in } X, \quad y(\cdot) \text{ in } Y. \qquad (1\text{-}1)$$

$T[\cdot]$ is called an input-output transformation.

Example

Consider the circuit shown in Figure 1.2. Let $x(\cdot)$ and $y(\cdot)$ be the input voltage and output voltage, respectively. Then a mathematical model for this system is the differential equation:

$$RC\,\frac{dy(t)}{dt} + y(t) \;=\; x(t)\;, \qquad t_0 < t < \infty,$$

where the input voltage is applied at some "initial" time t_0 (say).

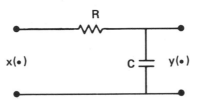

Figure 1.2.

To determine the input-output transformation $T[\cdot]$ in this case, we have to solve the differential equation for $y(\cdot)$. Multiplying both sides of the equation by $e^{\alpha t}$, $\alpha = \frac{1}{RC}$, we find

$$\frac{d}{dt}[e^{\alpha t}\,y(t)] \;=\; \alpha e^{\alpha t}\,x(t)\;, \qquad t \geq t_0\;.$$

Therefore

$$y(t) = K e^{-\alpha t} + \int_{t_0}^{t} \alpha e^{-\alpha(t-\sigma)} x(\sigma) \, d\sigma, \quad t \geq t_0,$$

where K is a constant to be determined. Setting $t = t_0$ we get

$$y(t_0) = K e^{-\alpha t_0}.$$

Therefore, $K = e^{\alpha t_0} y(t_0)$. Thus the output voltage is

$$y(t) = e^{-\alpha(t-t_0)} y(t_0) + \int_{t_0}^{t} \alpha e^{-\alpha(t-\sigma)} x(\sigma) \, d\sigma,$$

$$t \geq t_0$$

It is evident from this relation that the output voltage $y(\cdot)$ depends on the input voltage $x(\cdot)$ and the "initial" condition $y(t_0)$ -- the voltage across the capacitor C at the time t_0. Thus the same input $x(\cdot)$ causes many outputs $y(\cdot)$, depending on different values of $y(t_0)$. Now if we choose $y(t_0) = 0$ then clearly, for $t \geq t_0$:

$$y(t) = \int_{t_0}^{t} \alpha e^{-\alpha(t-\sigma)} x(\sigma) \, d\sigma.$$

We can now say that the circuit transforms each input voltage $x(\cdot)$ into a unique corresponding output voltage $y(\cdot) = T[x(\cdot)]$ -- provided there is no voltage (or, equivalently, no charge) across the capacitor at the time the input voltage was applied, and we write

4

$$y(\cdot) \;=\; T[x(\cdot)] \quad , \qquad y(t) \;=\; \int_{t_0}^{t} \alpha e^{-\alpha(t-\sigma)} \, x(\sigma) \, d\sigma \quad ,$$

$$t \geq t_0 \; .$$

Finally, for the same circuit, if the input is the current $i(\cdot)$, then the mathematical model is simply

$$y(t) \;=\; \frac{1}{C} \int_{t_0}^{t} i(\sigma) \, d\sigma \;+\; y(t_0) \quad , \qquad t \geq t_0 \; .$$

Therefore, as in the previous case, with $y(t_0) = 0$, we have the input-output transformation:

$$y(\cdot) \;=\; T[i(\cdot)] \quad , \qquad y(t) \;=\; \frac{1}{C} \int_{t_0}^{t} i(\sigma) \, d\sigma \quad , \qquad t \geq t_0 \; .$$

PROPERTIES OF AN INPUT-OUTPUT TRANSFORMATION

Properties of a system are now studied via those of its input-output transformation.

First we define.

Definition

A system with an input-output transformation $y(\cdot) = T[x(\cdot)]$ is said to be <u>linear</u> if

(i) for any scalar k and any $x(\cdot)$ in X:

$$T[kx(\cdot)] \;=\; kT[x(\cdot)] \; ;$$

(ii) for any $x_1(\cdot)$ and $x_2(\cdot)$ in X:

$$T[x_1(\cdot) + x_2(\cdot)] \;=\; T[x_1(\cdot)] + T[x_2(\cdot)] \; .$$

5

It is evident that (i) and (ii) can be combined into the single condition

$$T[k_1 x_1(\cdot) + k_2 x_2(\cdot)] = k_1 T[x_1(\cdot)] + k_2 T[x_2(\cdot)] \qquad (1.2)$$

for any scalars k_1, k_2 and any $x_1(\cdot)$, $x_2(\cdot)$ in X.

We must note that in the above we have assumed that $kx(\cdot)$ and $x_1(\cdot) + x_2(\cdot)$ are in X for each k and for any $x(\cdot)$, $x_1(\cdot)$ and $x_2(\cdot)$ in X. This is the same as saying that X is a linear space. Similarly Y is also a linear space, and the above Definition means that $T[\cdot]$ is a linear transformation from X to Y.

It follows at once from (1.2) that $T[0] = 0$. Thus a linear system -- in the sense of the above Definition -- must be such that a "zero" input (x=0) always results in a "zero" output (y=0). This implies that we are only concerned with the class of systems which are at "rest" -- i.e., zero input results in zero output -- at the time an input is applied to the systems.

When a system does not have a linear input-output transformation it is said to be nonlinear.

The next important property is that of a time-invariant (or fixed, or stationary) system. Heuristically speaking, a system is time-invariant if an input is shifted along the time axis by an amount τ (say), then the corresponding output is shifted by the same amount as illustrated in Figure 1.3. Clearly $x(t-\tau)$ for $\tau > 0$ (resp. $\tau < 0$) is just $x(t)$ shifted to the right (resp.

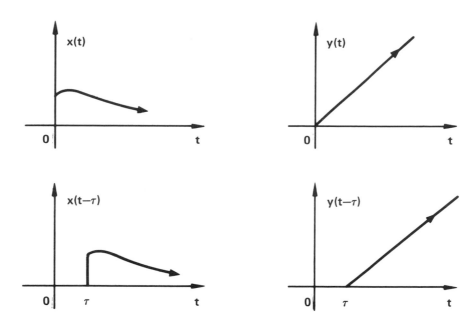

Figure 1.3.

left) by the amount τ. We therefore have the next Definition.

Definition

A system with an input-output transformation $y(\cdot) = T[x(\cdot)]$ is time-invariant if, for any t and any τ:

$$y(t) = T[x(t)]$$

and

$$z(t) = T[x(t-\tau)] \quad .$$

Then

$$z(t) = y(t-\tau) \quad .$$

Example

Consider

$$y(\cdot) \;=\; T[x(\cdot)] \quad , \quad y(t) \;=\; \int_{-\infty}^{\infty} (t-\sigma) \, x(\sigma) \, d\sigma \quad ,$$

$$-\infty < t < \infty \quad .$$

We have

$$T[x(t-\tau)] \;=\; \int_{-\infty}^{\infty} (t-\sigma) \, x(\sigma-\tau) \, d\sigma \;=\; z(t)$$

and

$$y(t-\tau) \;=\; \int_{-\infty}^{\infty} (t-\tau-\sigma) \, x(\sigma) \, d\sigma \quad .$$

Therefore $z(t) = y(t-\tau)$ and the system is time-invariant.

If a system is not time-invariant then it is called time-varying.

It is important to note that for a time-invariant system shifting an input along the time axis does not change the shape of the corresponding output. Therefore, for time-invariant systems, the origin of time can always be taken to be 0. In other words, if $x(\cdot)$ is applied to a time-invariant system at some time t_0 ($\neq 0$), then we can always shift it to the origin and take 0 to be the time at which the input is applied to the system. Consequently we can take $x(t)$ to be 0 for $t < 0$.

The next important concept is that of causality or physical realizability. A system is said to be causal if the value $y(t)$ at any time t of an output $y(\cdot)$ depends

8

only on the values of an input x(·) up to the time t
-- i.e., the values x(σ) for each σ ≤ t. Thus if we
regard t as the <u>present time</u> -- i.e., "now" -- then for
a causal system present value of an output can only depend
on past and present values of the input that causes it --
but not on the future values of the input.

If for a causal system and for any t, y(t) only
depends on x(t) -- i.e., present value of output only
depends on present value of input -- then the system is
called <u>instantaneous</u> or <u>memoryless</u> (or <u>zero memory</u>).
Otherwise it is said to have memory.

Example

The system with the input-output transformation

$$y(\cdot) \;=\; T[x(\cdot)] \;,\qquad y(t) \;=\; \int_{-\infty}^{t} x(\sigma)\, d\sigma \;,\qquad -\infty < t < \infty$$

is causal and has memory, while

$$y(t) \;=\; \int_{0}^{\infty} e^{-(t-\sigma)} x(\sigma)\, d\sigma \;,\qquad -\infty < t < \infty$$

is not causal. The system defined by

$$y(\cdot) \;=\; T[x(\cdot)] \;,\qquad y(t) \;=\; 4x(t) + 2$$

is memoryless.

9

Finally, if the time variable t -- of the input and
output signals -- takes all values in an interval (finite
or nonfinite) then the system is said to be continuous-
time. If t takes only discrete values then the system
is called discrete-time.

PROBLEMS

1. Verify whether the following input-output transforma-
    tions are: linear, nonlinear, time-invariant, time-
    varying, causal, noncausal, or memoryless:

$$y(t) \;=\; tx(t) \;,\qquad\qquad\qquad 0 \le t \le 5 \;;$$

$$y(t) \;=\; \int_{-\infty}^{t} e^{-(t-\sigma)}\sigma x(\sigma) \; d\sigma \;,\qquad\qquad -\infty < t < \infty \;;$$

$$y(t) \;=\; x(t) \;+\; \int_{0}^{t} (t-\tau)\, x(\tau)\; d\tau \;,\qquad t \ge 0 \;;$$

$$y(t) \;=\; \frac{dx(t)}{dt} \;-\; \int_{t}^{\infty} t\, \tau^2\, x(\tau)\; d\tau \;,\qquad t > 0 \;;$$

$$y(t) \;=\; 4x(t)2 \;,$$

$$y(t) \;=\; x(t-5) \;,\qquad\qquad\qquad -\infty < t < \infty \;;$$

$$y(t) \;=\; \int_{-\infty}^{\infty} t\, \sigma x(\sigma) \; d\sigma \;,\qquad\qquad -\infty < t < \infty \;,$$

2. (i) For S_1 and S_2 shown, show clearly whether they are: linear, time-varying and causal.

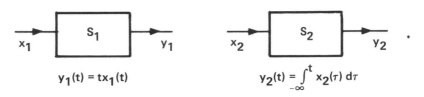

$$y_1(t) = t x_1(t)$$

$$y_2(t) = \int_{-\infty}^{t} x_2(\tau)\, d\tau$$

(ii) Find the input-output relation for the system S_3 formed by cascading S_1 and S_2 as shown.

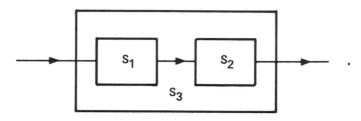

3. In problem 2.(ii) if S_1 and S_2 are interchanged, would the resulting system be the same as the system S_3?

4. (i) For systems S_1 and S_2 below, show clearly whether they are: linear or nonlinear, time-invariant or time-varying:

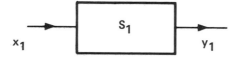

$$y_1(t) = \int_{0}^{t} \sigma x_1(\sigma)\, d\sigma , \qquad t \geq 0 .$$

4.   (i)   (continued)

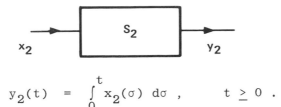

$$y_2(t) = \int_0^t x_2(\sigma) \, d\sigma \, , \qquad t \geq 0 \, .$$

(Note that both $x_1(t)$ and $x_2(t)$ are taken to be 0 for $t < 0$.)

(ii)   Obtain the input-output description for the system obtained by "cascading" S_1 and S_2 as shown below:

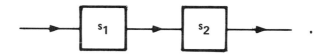

CHAPTER 2. LINEAR SYSTEMS: TIME-DOMAIN ANALYSIS

Analysis of linear systems is taken up in
this chapter. Given a system which can be
described by an abstract input-output transformation $y(\cdot) = T[x(\cdot)]$, our main concern here is to find, from the assumption
that $T[\cdot]$ is linear, an explicit relationship between an input and its corresponding
output. In other words, given a linear system, the analysis problem becomes that of
finding the output of the system when a given
input is applied to it. The key idea here is
the notion of <u>an impulse response function</u> of
a linear system.

THE DIRAC DELTA (OR IMPULSE) FUNCTION

The Dirac delta function, denoted by $\delta(\cdot)$ is defined as

$$\begin{cases} \delta(t) = 0 \qquad \text{for each } t \neq 0 , \\ \delta(0) = \infty , \\ \int_{-\infty}^{\infty} \delta(t)\, dt \;=\; 1 \end{cases}$$

and is represented by a pulse as shown in Figure 2.1.

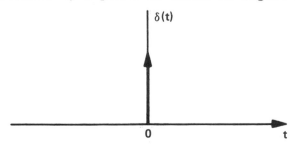

Figure 2.1.

13

Strictly speaking, the delta function is <u>not</u> a function in the usual sense of the word. However one can think of it as the limit of a rectangular pulse $p_w(\cdot)$ of width w and of height $\frac{1}{w}$ as shown in Figure 2.2.

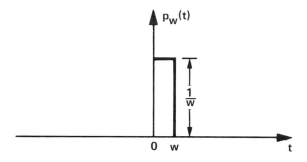

Figure 2.2.

It is clear that

$$p_w(t) \;=\; 0 \qquad \text{for } t > w \text{ and for } t < 0 \;,$$

and

$$\int_{-\infty}^{\infty} p_w(t) \; dt \;=\; w \cdot \frac{1}{w} \;=\; 1 \;.$$

Moreover

$$p_w(t) \;=\; \frac{1}{w} \qquad \text{for } 0 \le t \le w \;.$$

Therefore

$$\delta(t) \;=\; \lim_{w \to 0} p_w(t) \;,$$

as expected.

Let $f(\cdot)$ be a given function defined for $-\infty < t < \infty$, then it is clear that

$$f(t)\; \delta(t) \;=\; f(0)\; \delta(t) \;.$$

14

Then

$$\int_{-\infty}^{\infty} f(t)\ \delta(t)\ dt\ =\ \int_{-\infty}^{\infty} f(0)\ \delta(t)\ dt$$

$$=\ f(0) \int_{-\infty}^{\infty} \delta(t)\ dt\ =\ f(0)\ .$$

Thus we can think of the effect of the delta function in

the integral $\int_{-\infty}^{\infty} f(t)\delta(t)\ dt$ as to "pick" out the value

of the function $f(\cdot)$ at 0. Similarly, for any t and

σ:

$$f(t)\ \delta(t-\sigma)\ =\ f(\sigma)\ \delta(t-\sigma)\ \ ,$$

and therefore

$$\int_{-\infty}^{\infty} f(t)\ \delta(t-\sigma)\ dt\ =\ f(\sigma)\ . \qquad (2.1)$$

Interchanging t and δ we find

$$f(t)\ =\ \int_{-\infty}^{\infty} f(\sigma)\ \delta(\sigma-t)\ d\sigma\ . \qquad (2.2)$$

This integral is just the resolution of $f(\cdot)$ into a
continuum of delta functions. To see this, we approximate
$f(\cdot)$ in a finite interval $[-a,a]$ (say) by a number of
rectangular pulses $p_w(\cdot)$ shown in Figure 2.3. It follows
that, for $-a \le t \le a$:

$$f(t)\ \underset{\sim}{}\ \sum_{-n}^{n} f(kw)\ p_w(t-kw)\ w\ . \qquad (2.3)$$

15

This then becomes, as $w \to 0$ and $n \to \infty$:

$$f(t) \;=\; \int_{-a}^{a} f(\sigma)\, \delta(t-\sigma)\, d\sigma \quad .$$

Letting a go to ∞ we obtain (2.2) above.

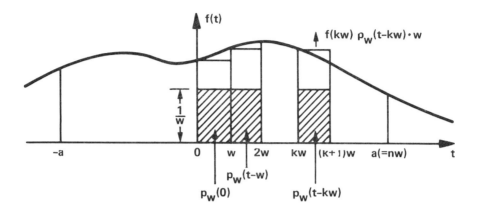

Figure 2.3.

THE UNIT STEP FUNCTION

The unit step function is denoted by $U(\cdot)$ and is defined as, for $-\infty < t < \infty$:

$$U(t) \;=\; \begin{cases} 1\,, & \text{for } t \geq 0 \\ 0\,, & \text{for } t < 0 \end{cases} \quad .$$

It follows at once that

$$U(-t) \;=\; \begin{cases} 1\,, & \text{for } t \leq 0 \\ 0\,, & \text{for } t > 0 \end{cases} \quad .$$

16

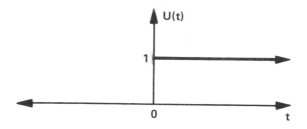

Figure 2.4a.

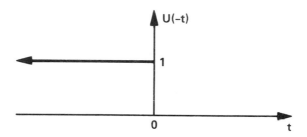

Figure 2.4b.

The unit step function is often used to "truncate" other functions. Thus, given $f(\cdot)$ for $-\infty < t < \infty$, then for any fixed τ:

$$f(t) \, U(t-\tau) \;=\; \begin{cases} f(t) \,, & \text{for} \quad t-\tau \geq 0 \\ 0 \,, & \text{for} \quad t-\tau < 0 \,. \end{cases}$$

Therefore the function $f(\cdot)$ has been truncated at $t = \tau$ by the unit step function $U(t-\tau)$.

It is evident from the definitions of $\delta(\cdot)$ and $U(\cdot)$ that

17

$$\delta(t) = \frac{dU(t)}{dt}$$

and

$$U(t) = \int_{-\infty}^{t} \delta(\tau) \, d\tau \quad .$$

THE IMPULSE RESPONSE FUNCTION OF A LINEAR SYSTEM

The impulse function $\delta(t-\tau)$ for $-\infty < t, \quad \tau < \infty$, as we know, is located at τ on the t-axis -- if t is the independent variable.

Definition

The response of a linear system with an input-output transformation $y(\cdot) = T[x(\cdot)]$ to the input $\delta(t-\tau)$ is called the <u>impulse response function</u>, and is denoted by $h(t,\tau) \ (= T[\delta(t-\tau)])$.

Plainly speaking, $h(t,\tau)$ is just the response of the system -- <u>at time t</u> -- due to an impulse δ -- <u>applied at time τ</u>.

Example

For the system with the input-output transformation

$$y(t) = \int_{0}^{\infty} e^{-(t-\sigma)} U(t-\sigma) \, x(\sigma) \, d\sigma \ , \qquad t \geq 0 \ ,$$

we have

$$h(t,\tau) = \int_0^\infty e^{-(t-\sigma)} U(t-\sigma) \delta(\sigma-\tau) d\sigma \quad ,$$

$$= e^{-(t-\tau)} U(t-\tau) \quad .$$

We note that $h(t,\tau)$ in this case is a function of $(t-\tau)$, therefore we can write $h(t,\tau) = h(t-\tau)$ $= e^{-(t-\tau)}U(t-\tau)$. It then follows that $h(t,0) = h(t)$ $= e^{-t}U(t)$.

Next, for the input-output relation:

$$y(t) = \int_{-\infty}^\infty \sigma t \, U(t-\sigma) \, x(\sigma) \, d\sigma$$

we find

$$h(t,\tau) = \tau t \, U(t-\tau)$$

which, unlike the previous case, is not a function of $(t-\tau)$.

Now suppose that the system is time-invariant. Then, since $\delta(t-\tau)$ is just $\delta(t)$ shifted by an amount τ, we have

$$\left.\begin{array}{l} h(t,0) = T[\delta(t)] \\ \text{and} \\ h(t,\tau) = T[\delta(t-\tau)] \end{array}\right\} \quad => \quad h(t,\tau) = h(t-\tau,0) \quad .$$

Thus the impulse response function $h(t,\tau)$ of a linear time-invariant system depends on $(t-\tau)$ -- but not on t and τ separately. We shall write $h(t-\tau)$ -- in place of $h(t-\tau,0)$ -- or sometimes just $h(t)$.

Finally, if the system is <u>causal</u> then it must follow

that $\underline{h(t,\tau) = 0}$ for $t < \tau$, otherwise the output $h(t,\tau)$ at time $t < \tau$ would be the response due to the input $\delta(t-\tau)$ at a future time τ. This, of course, is not possible since the system is causal.

We note that for a noncausal system, a nonzero input can result in a zero output as in the next example.

Example

Consider the input-output relation:

$$y(t) \;=\; x(t) \;-\; \int_t^\infty 2e^{-(\tau-t)} x(\tau) \, d\tau \;, \qquad t \geq 0$$

which represents a noncausal system since the output $y(t)$ depends on all values of the input $x(\tau)$ for all $t \leq \tau < \infty$. Setting $y(t) = 0$ we obtain

$$x(t) \;=\; 2 \int_t^\infty e^{-(\tau-t)} x(\tau) \, d\tau \;, \qquad t \geq 0 \quad .$$

Therefore

$$\frac{dx}{dt} \;=\; 2 \int_t^\infty e^{-(\tau-t)} x(\tau) \, d\tau \;-\; 2x(t) \;,$$

$$\;=\; x(t) - 2x(t)$$

$$\;=\; -x(t) \;, \qquad\qquad t \geq 0 \quad .$$

From which it follows that any input of the form

$$x(t) \;=\; Ke^{-t} \;, \qquad t \geq 0$$

would result in a zero output.

INPUT-OUTPUT RELATION OF A LINEAR SYSTEM

Consider a linear system with a given input $x(\cdot)$. Then the corresponding output $y(\cdot)$ is, at least formally: $y(\cdot) = T[x(\cdot)]$. Here we are concerned with the problem of finding an explicit relationship between the inputs and outputs of a linear system.

First an input $x(t)$ can be formally written as, for each t in $(-\infty, \infty)$:

$$x(t) = \int_{-\infty}^{\infty} \delta(t-\tau) \, x(\tau) \, d\tau \ .$$

Therefore it follows that

$$y(t) = T[x(t)] = T\left[\int_{-\infty}^{\infty} \delta(t-\tau) \, x(\tau) \, d\tau\right], \quad -\infty < t < \infty \ .$$

Then, since the system is linear, we can interchange $T[\cdot]$ and the integration to obtain

$$y(t) = \int_{-\infty}^{\infty} T[\delta(t-\tau)] \, x(\tau) \, d\tau, \quad -\infty < t < \infty \ .$$

Therefore

$$y(t) = \int_{-\infty}^{\infty} h(t,\tau) \, x(\tau) \, d\tau, \quad -\infty < t < \infty$$

by the fact that $T[\delta(t-\tau)] = h(t,\tau)$. Thus we have shown the following theorem.

21

Theorem

Let $h(t, \tau)$ be the impulse response function of a linear system. Then an input $x(\cdot)$ to the system and its corresponding output $y(\cdot)$ are related by the <u>superposition integral</u>:

$$y(t) = \int_{-\infty}^{\infty} h(t, \tau) \, x(\tau) \, d\tau \, , \qquad -\infty < t < \infty \, . \quad (2.4)$$

In addition, if the system is also time-invariant, then

$$y(t) = \int_{-\infty}^{\infty} h(t-\tau) \, x(\tau) \, d\tau = \int_{-\infty}^{\infty} h(\tau) \, x(t-\tau) \, d\tau \, ,$$

$$-\infty < t < \infty \, . \quad (2.5)$$

Example

The input $x(\cdot)$ and output $y(\cdot)$ of a linear system are related by the differential equation

$$\frac{dy(t)}{dt} + a(t) \, y(t) = x(t) \, , \qquad t > 0 \, ,$$

where $a(\cdot)$ is a given function. To find the input-output transformation we solve the equation for $y(\cdot)$.

First let us take $t = 0$ to be the time at which input was applied to the system. Furthermore, since the system is linear, we have $y(\cdot) = T[0] = 0$. Therefore if $x(t) = 0$ for $t < 0$, then $y(t) = 0$ for $t < 0$ also. At $t = 0$ we also take $y(0) = 0$ since up to that time no input was applied to the system, hence it must be at rest. Thus we solve the above differential equation sub-

22

ject to the initial condition

$$y(0) \;=\; 0 \;.$$

For this we multiply both sides of the equation by the "integrating factor"

$$e^{\left(\int_0^t a(\sigma)\,d\sigma\right)}$$

and obtain

$$\frac{d}{dt}\left[e^{\left(\int_0^t a(\sigma)d\sigma\right)}y(t)\right] \;=\; e^{\left(\int_0^t a(\sigma)d\sigma\right)}x(t) \;\;.$$

Integrating both sides from 0 to t we get

$$e^{\left(\int_0^t a(\sigma)d\sigma\right)}y(t) \;=\; \int_0^t e^{\left(\int_0^\tau a(\sigma)d\sigma\right)}x(\tau)\;d\tau\,,$$

where, on the left-hand side, we have made use of the fact that $y(0) = 0$. Finally

$$y(t) \;=\; \int_0^t e^{\left(\int_t^\tau a(\sigma)d\sigma\right)}x(\tau)\;d\tau\;, \qquad t > 0\,,$$

which is the desired input-output relation.

It follows easily that the impulse response function of the system is

$$h(t,\tau) \;=\; e^{\left(\int_t^\tau a(\sigma)d\sigma\right)}U(t-\tau) \;\;.$$

Thus the system is physically realizable. Note that if

$a(\cdot)$ is a constant function α (say), then

$$h(t,\tau) \;=\; e^{-\alpha(t-\tau)}\; U(t-\tau) \quad,$$

and the system is therefore time-invariant in this case
as expected.

   The superposition integral (2.4) above is, in general,
the input-output transformation of a <u>noncausal</u> system
since the output at time t depends on all vaues $x(\tau)$
for $-\infty < \tau < t$ and for $t \le \tau < \infty$.   Thus, if the system
is causal then we must have $h(t,\tau) = 0$,   for $t < \tau$.
Consequently (2.4) becomes:

$$y(t) \;=\; \int_{-\infty}^{\infty} h(t,\tau)\; U(t-\tau)\; x(\tau)\; d\tau \quad,$$

$$ \;=\; \int_{-\infty}^{t} h(t,\tau)\; x(\tau)\; d\tau \quad.$$

Now if the system is causal and time-invariant then it
follows that

$$y(t) \;=\; \int_{-\infty}^{t} h(t-\tau)\; x(\tau)\; d\tau \quad, \qquad t > -\infty \quad.$$

Moreover, we have seen that for a time-invariant system
we can always choose 0 to be the time origin.   This allows
us, for a linear time-invariant system, to choose $x(t) = 0$
for $t < 0$.   We therefore have, for a linear time-invariant and causal system with inputs $x(t)U(t)$:

24

$$y(t) \quad = \quad \int_{-\infty}^{t} h(t-\tau) \; x(\tau) \; U(\tau) \; d\tau \quad , \qquad t \geq 0 \quad ,$$

$$= \quad \int_{0}^{t} h(t-\tau) \; x(\tau) \; d\tau \; , \qquad t \geq 0 \quad .$$

We summarize the above in:

Corollary

A linear time-invariant and causal system has the input-output transformation, for any t:

$$y(t) \quad = \quad \int_{-\infty}^{t} h(t-\tau) \; x(\tau) \; d\tau \quad = \quad \int_{0}^{\infty} h(\tau) \; x(t-\tau) \; d\tau \quad . \qquad (2.6)$$

If in addition $x(t) = 0$ for $t < 0$ then, for $t \geq 0$:

$$y(t) \quad = \quad \int_{0}^{t} h(t-\tau) \; x(\tau) \; d\tau \quad = \quad \int_{0}^{t} h(\tau) \; x(t-\tau) \; d\tau \qquad (2.7)$$

which is called a convolution integral.

We note from (2.6) that for a linear time-invariant and causal system, even though $h(t) = 0$ for $t < 0$, its inputs $x(t)$ can be nonzero for $t < 0$.

IMPULSE RESPONSE FUNCTION OF A CASCADED SYSTEM

Let S_1 and S_2 be two linear systems with impulse response functions $h_1(t,\tau)$ and $h_2(t,\tau)$, respectively. We now wish to find the impulse response function of the system S formed by cascading S_1 and S_2 as shown in Figure 2.5.

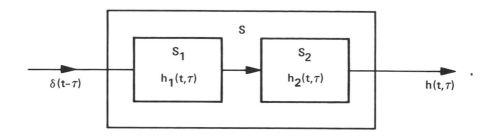

Figure 2.5.

It is evident that, if $h_{12}(t,\tau)$ is the impulse response function of S then

$$h_{12}(t,\tau) = \int_{-\infty}^{\infty} h_2(t,\sigma)\, h_1(\sigma,\tau)\, d\sigma \quad .$$

Note that, in general, cascading S_1 and S_2 is not the same as cascading S_2 and S_1, except when the two systems are time-invariant. Indeed, if this is the case, then for the cascaded system "$S_1 S_2$" we have

$$h_{12}(t-\tau) = \int_{-\infty}^{\infty} h_2(t-\sigma)\, h_1(\sigma-\tau)\, d\sigma \quad .$$

Setting $t-\sigma = \xi$ (say) we find

$$h_{12}(t-\tau) = \int_{-\infty}^{\infty} h_1(t-\xi-\tau)\, h_2(\xi)\, d\xi \quad ,$$

or

$$h_{12}(t-\tau) = \int_{-\infty}^{\infty} h_1(t-\lambda)\, h_2(\lambda-\tau)\, d\lambda \quad .$$

The integral on the right-hand side is clearly the impulse
response function $h_{21}(t-\tau)$ of the cascaded system
"S_2S_1". Therefore $h_{12}(t-\tau) = h_{21}(t-\tau)$ as expected.

PROBLEMS

1. Write down an expression for the function $f(\cdot)$
   shown below

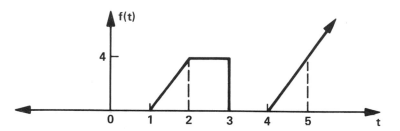

   Sketch the following functions:

$$f(t) = U(t-4)\, U(\sigma-t)\, , \qquad -\infty < t\, , \quad \sigma < \infty$$

$$g(t) = U(3-t)\, u(t+1)\, , \qquad -\infty < t < \infty\, .$$

2. Given

$$h(t) = t\, e^{-t}\, U(t)\, ,$$

$$x(t) = 2[\delta(t) - t^2 U(t)]$$

   show that

$$\int_{-\infty}^{\infty} h(t-\tau)\, x(\tau)\, d\tau = \int_{-\infty}^{\infty} h(\tau)\, x(t-\tau)\, d\tau\, .$$

3. (i) Sketch $U(t-5)$ and $U(t+4)$, $-\infty < t < \infty$.

   (ii) Sketch $U(t\pm\sigma)$ for $-\infty < t < +\infty$, $\sigma > 0$ and
   $\sigma < 0$.

4. Sketch the following functions :

    (a) $f(t)$ = $U(t-2) U(\sigma-t)$;

    (b) $g(t)$ = $U(-t)(t+1)$;

    (c) write an expression for the function $h(\cdot)$ shown.

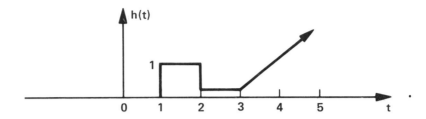

5. Compute the following integrals :

$$\int_{-\infty}^{\infty} e^{-t} t\, U(t-5)\, dt ;$$

$$\int_{-\infty}^{\infty} e^{+\tau} U(4-\tau)\, d\tau .$$

6. Show that

$$\int_{-\infty}^{\infty} \left[\delta(t-\sigma) - 2e^{-(t-\sigma)}U(t-\sigma)\right]\left[\delta(\sigma-\tau) - 2e^{-(\tau-\sigma)}U(\tau-\sigma)\right] d\sigma$$

$$= \delta(t-\tau) \qquad \text{for} \quad -\infty < t , \quad \tau < +\infty .$$

7. Compute

$$\int_{-\infty}^{\infty} e^{-2t} t^2 U(t+2) \, dt \; ;$$

$$\int_{-\infty}^{\infty} \left[\delta(t-\sigma) - e^{-(t-\sigma)} U(t-\sigma) \right] \left[\delta(\tau-\sigma) - e^{-(\tau-\sigma)} U(\tau-\sigma) \right] d\sigma \; ,$$

$$-\infty < t \; , \quad \tau < \infty \quad ;$$

$$\int_{-\infty}^{\infty} (t-\sigma) \, e^{-(t-\sigma)} U(t-\sigma) \, \sigma^2 U(\sigma) \, d\sigma \; , \quad t \geq 0 \; .$$

8. Find the impulse response function of the system whose input $x(\cdot)$ and output $y(\cdot)$ are related by the differential equation

$$\frac{dy}{dt} + ay(t) = x(t) \; , \quad t > 0 \; ,$$

where a is a constant, by directly solving the equation with $x(t) = \delta(t)$.

9. Let $h(t,\sigma)$ be the impulse response function, and $g(t,\sigma)$ be the unit step response function -- i.e, $g(t,\sigma) = T[U(t-\sigma)]$, of a linear system. Find a relationship between $h(t,\sigma)$ and $g(t,\sigma)$ and find an input-output transformation for the system using $g(t,\sigma)$.

10. Find the impulse response function of the system S_3 of Problem 2.(ii) of Chapter 1.

11. Obtain the resolution of a function $f(\cdot)$ into a continuum of unit step functions -- similar to the resolution (2.2).

12. Find the ouptut $y(t)$ of a linear time-invariant system whose impulse response function
$h(t) = te^{-t}U(t)$, and whose input $x(t) = e^{-2|t|}$,
$-\infty < t < \infty$.

13. A linear system is described by the differential equation
$$\frac{dy}{dt} + \frac{2t}{t^2+1} y(t) = x(t) , \qquad t > 0$$
and
$$y(0) = 0 .$$

Find the system impulse response function $h(t,\tau)$ and show that it is time-varying and causal.

14. Consider:

$$y(t) = \int_{-\infty}^{\infty} h(t-\tau) x(\tau) d\tau , \qquad \text{for} \quad -\infty < t < \infty .$$

Compute $y(t)$ given:
  (i) $h(t) = e^{-2t} U(t)$, $\quad x(t) = tU(t)$.
  (ii) $h(t) = t^2 U(t+1)$, $\quad x(t) = e^{-t}U(t+2)$.
  (iii) $h(t) = \delta(t) - tU(t)$.

14. (iii) (continued)

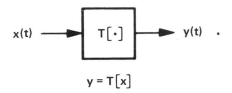

$$x(t) \quad = \quad$$

(iv) $h(t) \quad = \quad \delta(t) - e^{-t}U(t),$

$x(t) \quad = \quad \delta(t) - te^{-t}U(t).$

15. Find the impulse response function $h(t,\tau)$ of the following systems:

$$x(t) \longrightarrow \boxed{T[\cdot]} \longrightarrow y(t) \quad .$$

$$y = T[x]$$

(i) $y(t) \quad = \quad \displaystyle\int_{-\infty}^{t} U(\sigma) \ x(\sigma) \ d\sigma \ , \qquad -\infty < t < \infty \ .$

(ii) $y(t) \quad = \quad \displaystyle\int_{-\infty}^{\infty} tU(t-\sigma) \ x(\sigma) \ d\sigma \ , \qquad -\infty < t < \infty \ .$

(iii) $x(t) \longrightarrow \boxed{(1)} \longrightarrow z(t) \longrightarrow \boxed{(2)} \longrightarrow y(t)$

$$z(t) \quad = \quad e^{-t}x(t) \ U(t),$$

$$y(t) \quad = \quad \displaystyle\int_{0}^{t} e^{-(t-\sigma)} \ z(\sigma) \ U(\sigma) \ d\sigma \ , \quad t \geq 0 \ .$$

(iv) Find the impulse response function of the system obtained by interchanging system (1) and system (2) of part (iii).

31

16. Let S be a linear, time-invariant and causal system
    whose input x(t) and corresponding output y(t) are
    shown below:

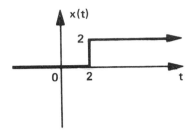

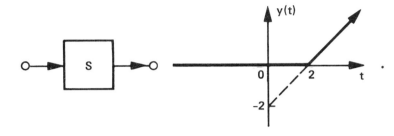

    (i) Find the impulse response function h(t) of S.
    (ii) Find the output of S when its input is

$$x(t) = \begin{cases} e^t & \text{for } t < 0 \\ e^{-t} & \text{for } t \geq 0 \end{cases}.$$

17. In the following h(t,τ) is the impulse response
    function of a linear system and x(t) is its input.
    Find the corresponding output.

    (i) $h(t,\tau) = t\,\tau^2\,U(t-\tau),$ $x(t) = e^{-t}U(t+1).$

17. (continued)

   (ii) $h(t,\tau) = \sin(t-\tau)\, U(t-\tau)$,

        $x(t) = (t-1)\, U(t-1)$.

   (iii) $h(t,\tau) = te^{\tau}$, $x(t) = U(5-t)\, \cos t\, U(t+3)$.

CHAPTER 3. LINEAR TIME-INVARIANT AND CAUSAL SYSTEMS:

LAPLACE TRANSFORM ANALYSIS

This chapter is devoted to the theory of the
Laplace Transform and its applications to the
analysis of linear time-invariant and causal
systems. The important notion of a system function is introduced and its relation to the
impulse response function is discussed.

THE LAPLACE TRANSFORM

DEFINITION

In the following we will be dealing with functions of
the time variable t as well as functions of a complex
variable s. The real and imaginary parts of a complex
quantity are denoted by Re. and Im., respectively.

Let f(t) be a function of the real variable t on
$[0,\infty)$. The Laplace Transform of f(t), denoted by
$L_s[f(t)]$, is defined as

$$L_s[f(t)] = \int_0^\infty e^{-st} f(t)\, dt \quad , \qquad (3.1)$$

as long as the integral exists.

It is clear that if the integral exists then it is a
function of the complex variable s. Therefore we write
F(s) for the Laplace Transform of a function f(t). Although s is complex, it is convenient for the moment to
think of it as being real, and it is "large enough"

34

for the integral to exist.

The Laplace Transform as defined is called the <u>one-side</u>
Laplace Transform since the function f(t) in (3.1) is being
integrated only from 0 to ∞. Thus we can consider f(t)
to be 0 for t < 0. Such a function can be regarded as
the impulse response function of a linear time-invariant
and causal system, and as a consequence it can be called a
<u>causal function</u> or <u>signal</u>. In this chapter we are dealing
with causal functions.

Example

We have, for $f(t) = e^{-t}$:

$$L_s[e^{-t}] = \int_0^\infty e^{-st} e^{-t} dt = \frac{-e^{-(s+1)t}}{(s+1)} \Big|_0^\infty .$$

Set $s = \alpha + i\omega$, then

$$e^{-(s+1)t} = e^{-(\alpha+1)t} e^{-i\omega t} .$$

Therefore

$$\left|e^{-(s+1)t}\right| = \left|e^{-(\alpha+1)t}\right| \times \left|e^{i\omega t}\right| = \left|e^{-(\alpha+1)t}\right| .$$

It then follows that if $\alpha + 1 > 0$ then $e^{-(\alpha+1)t}$ goes
to 0 as t goes to ∞. We therefore have

$$L_s[e^{-t}] = F(s) = \frac{1}{s+1} , \quad \text{for Re. } s > -1 .$$

Note that in this case the function F(s) "blows up" at
s = -1, hence -1 is called a <u>pole</u> of F(s). We have

35

the following "plot" for the function F(s) in the [s-plane] in Figure 3.1. The half-plane Re. s > -1 is called the domain of convergence of the Laplace Transform of e^{-t}.

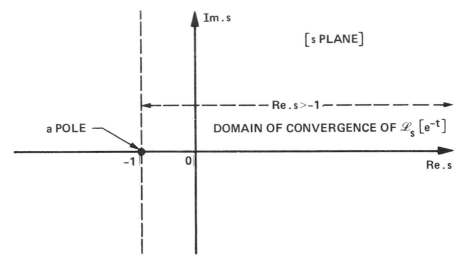

Figure 3.1.

SOME IMPORTANT TRANSFORMS

Listed below are the transforms of some elementary functions. It is important to note that, for our purpose, these transforms together with some basic properties -- to be discussed in the next section -- will be enough for us to find the transforms of most functions. Here a is complex while ω and b are real, also we have not specified the domain of convergence of each transform. This is left as a simple exercise.

f(t)	F(s)
$U(t)$	$\dfrac{1}{s}$
$\delta(t)$	1
e^{at}	$\dfrac{1}{s-a}$
$\sin \omega t$	$\dfrac{\omega}{s^2 + \omega^2}$
$\cos \omega t$	$\dfrac{s}{s^2 + \omega^2}$
$\sinh bt$	$\dfrac{b}{s^2 - b^2}$
$\cosh bt$	$\dfrac{s}{s^2 - b^2}$
$t^n, \quad n \geq 0$	$\dfrac{n!}{s^{(n+1)}}$

BASIC PROPERTIES

(i) <u>Linearity</u>. $L_s[\ ]$ is a linear transformation, that is, for any scalars a and b:

$$L_s[af_1(t) + bf_2(t)] \ = \ aL_s[f_1(t)] + bL_s[f_2(t)] \quad ,$$

$$= \ aF_1(s) + bF_2(s) \quad .$$

This is self-evident from the Definition of the Laplace Transform.

The linearity property, as we shall see, is one of the main properties that make Laplace Transforms work in the analysis of linear time-invariant systems. One can also use this property for computing Laplace Transforms.

Example

We have, since $\sin \omega t = \dfrac{e^{i\omega t} - e^{-i\omega t}}{2i}$,

$$L_s[\sin \omega t] = \frac{1}{2i} L_s[e^{i\omega t}] - \frac{1}{2i} L_s[e^{-i\omega t}] ,$$

$$= \frac{1}{2i} \left\{ \frac{1}{s - i\omega} - \frac{1}{s + i\omega} \right\} ,$$

$$= \frac{\omega}{s^2 + \omega^2} .$$

(ii) If $F(s) = L_s[f(t)]$ then, for any a -- real or complex:

$$L_s[e^{-at} f(t)] = F(s+a) . \qquad (3.2)$$

This again follows immediately from the Definition of the Laplace Transform. It is clear that one can use this result to find the transform of any function of the form $e^{-at}f(t)$ once the transform $F(s)$ of $f(t)$ is known.

Example

For any integer n (≥ 0), since $L_s[t^n] = \dfrac{n!}{s^{n+1}}$,

$$L_s[e^{-\alpha t} t^n] = \frac{n!}{(s+\alpha)^{(n+1)}} .$$

(iii) <u>Transforms of derivatives of a function.</u> Let F(s) be the Laplace Transform of f(t), we now wish to find $L_s\left[\dfrac{df}{dt}\right]$. We have, by definition

$$L_s\left[\frac{df}{dt}\right] = \int_0^\infty e^{-st} \frac{df}{dt}\, dt \quad .$$

Therefore, by integration by parts

$$L_s\left[\frac{df}{dt}\right] = f(t)\, e^{-st}\Big|_0^\infty + s\int_0^\infty e^{-st}\, f(t)\, dt \quad ,$$

$$= \lim_{t\to\infty} f(t)\, e^{-st} - f(0) + sF(s) \quad .$$

Thus if the function f(t) is such that, for each s for which F(s) exists:

$$\lim_{t\to\infty} f(t)\, e^{-st} = 0 \quad ,$$

then

$$L_s\left[\frac{df}{dt}\right] = sL_s[f(t)] - f(0) \quad . \qquad (3.3)$$

It is evident from this that <u>if f(0) = 0, then multipli-cation by s -- in the complex domain -- corresponds to $\dfrac{d}{dt}$ in the time-domain.</u>

In exactly the same way we obtain

$$L_s\left[\frac{d^2 f}{dt^2}\right] = s^2 L_s[f(t)] - sf(0) - f'(0) \quad . \qquad (3.4)$$

In general, for any integer $n \geq 0$:

$$L_s\left[\frac{d^n f}{dt^n}\right] = s^n L_s[f(t)] - s^{n-1}f(0) - s^{n-2}f'(0)$$

$$- \cdots - f^{(n-1)}(0) \quad . \qquad (3.5)$$

(iv) <u>Transform of $\int_0^t f(\tau)d\tau$</u>. If $F(s)$ is the transform of $f(t)$, then

$$L_s\left[\int_0^t f(\tau) \, d\tau\right] = \frac{F(s)}{s} \quad . \qquad (3.6)$$

This again follows readily from the Definition of $L_s[\ ]$.

(v) <u>Initial Value Properties</u>. Consider $L_s[\delta(t)]$. We know that the delta function is not defined at the origin -- since $\delta(0) = \infty$. Therefore for $L_s[\delta(t)]$ to make sense we define 0^- to be $\lim(-\varepsilon)$ as ε goes to 0 and take

$$L_s[\delta(t)] = \int_{0^-}^{\infty} e^{-st} \delta(t) \, dt \quad .$$

Thus the delta function "sits" in the half-line $[0^-, \infty)$; as a consequence, its Laplace Transform is just the constant function 1.

In the following we also need 0^+ which is defined as $\lim(+\varepsilon)$ as ε goes to 0. Note that we can, of course, move the delta function to 0^+, i.e., take $\delta(0) = 0$ and $\delta(0^+) = \infty$, to obtain $L_s[\delta(t)] = 1$ -- without changing the definition of the Laplace Transform.

We have shown above that $L_s[f'(t)] = sF(s) - f(0)$. Thus if the function $f(\cdot)$ has discontinuity at $t = 0$ then $f(0)$ has to be carefully interpreted.

We either have $f(0) = f(0^-)$ or $f(0) = f(0^+)$ and

$$\int_{0^-}^{\infty} e^{-st} f'(t) \, dt = sF(s) - f(0^-) \, , \qquad (3.7)$$

or

$$\int_{0^+}^{\infty} e^{-st} f'(t) \, dt = sF(s) - f(0^+) \, . \qquad (3.8)$$

Next, as s becomes large, $s \to \infty$, the left-hand side of (3.8) goes to 0, then

$$f(0^+) = \lim_{s \to \infty} sF(s) \, . \qquad (3.9)$$

This result allows us to obtain the initial value $f(0^+)$ of $f(t)$ from its Laplace Transform, and is called the initial value theorem.

Applying (3.5) and the above discussion to the delta function we find, for $n = 1, 2, \ldots$:

$$L_s\left[\frac{d^n \delta(t)}{dt^n}\right] = s^n \, . \qquad (3.10)$$

(vi) Laplace Transform of a Delayed Signal. Given $f(t)$, $0 \leq t \leq \infty$, then for any fixed t_d (> 0) $f(t-t_d)$ is just $f(t)$ "delayed" by an amount t_d. We have

41

$$L_s[f(t-t_d)] = \int_0^\infty e^{-st} f(t-t_d) \, dt \quad,$$

$$= \int_{t_d}^\infty e^{-st} f(t-t_d) \, dt \quad,$$

where it is understood that $f(t)$ is 0 for $t < 0$, i.e., we could have written $f(t)U(t)$. Setting $t-t_d = \xi$ (say) we find

$$L_s[f(t-t_d)] = \int_0^\infty e^{-s(\xi+t_d)} f(\xi) \, d\xi \quad.$$

Therefore, for $t_d > 0$:

$$L_s[f(t-t_d)] = e^{-st_d} F(s) \quad, \qquad (3.11)$$

where $F(s)$ is $L_s[f(t)]$.

INVERSE TRANSFORMS: GIVEN $F(s) = L_s[f(t)]$ FIND $f(t)$

If $F(s)$ is the Laplace Transform of $f(t)$ then $f(t)$ is called the inverse (Laplace) transform of $F(s)$. Thus the inversion problem is to find $f(t)$ whose transform $F(s)$ is given.

Here we shall concentrate only on the class of $F(s)$'s which are rational functions of s, that is, of the form

$$F(s) = \frac{P(s)}{Q(s)} \quad,$$

where P and Q are polynomials in s.

The basic steps in finding $f(t)$ from a given $F(s)$ are:

<u>STEP 1</u>

If <u>deg. P(s) $\geq$ deg. Q(s)</u>: divide $P(s)$ by $Q(s)$ to obtain

$$F(s) = R(s) + \frac{\tilde{P}(s)}{Q(s)} \, ,$$

where, of course, deg. $\tilde{P}(s) <$ deg. $Q(s)$.

<u>Example</u>

$$F(s) = \frac{s-1}{s+1} \, .$$

We have, in this case, deg. $P(s) =$ deg. $Q(s) = 1$, therefore on dividing,

$$F(s) = 1 - \frac{2}{s+1} \, .$$

Hence it is clear that the inverse transform $f(t)$ of $F(s)$ is

$$f(t) = \delta(t) - 2e^{-t} U(t) \, .$$

<u>STEP 2</u>

Let <u>deg. P(s) $<$ deg. Q(s)</u> and suppose that $\underline{Q(s)}$ <u>is of degree</u> n and has n <u>distinct</u> roots $a_1, a_2, \ldots, a_n$ (say). Then expand $F(s)$ into <u>partial fractions</u>, i.e.,

$$F(s) = \frac{P(s)}{Q(s)} = \frac{P(s)}{(s-a_1)(s-a_2)\cdots(s-a_n)} \, ,$$

$$= \frac{A_1}{(s-a_1)} + \frac{A_2}{(s-a_2)} + \cdots + \frac{A_n}{(s-a_n)} \, ,$$

where, clearly,

$$A_r = \left. \left| \frac{P(s)(s-a_r)}{Q(s)} \right| \right|_{s=a_r} \quad , \quad r = 1, 2, \ldots, n \quad .$$

Therefore it follows that

$$f(t) = \sum_{r=1}^{n} A_r e^{a_r t} \quad .$$

Note that the roots a_r can of course be real or complex.

STEP 3

Let deg. P(s) < deg. Q(s) = n and suppose that Q(s) has repeated roots. For instance, let a_2 be a double root of Q(s), then of course Q(s) contains the factor $(s-a_2)^2$. In the partial fractions expansion of F(s) we have, corresponding to the factor $(s-a_2)^2$:

$$\frac{A_2}{(s-a_2)} + \frac{B_2}{(s-a_2)^2} \quad .$$

Here B_2 is given, as in the previous case, by

$$B_2 = \left. \left| \frac{P(s)(s-a_2)^2}{Q(s)} \right| \right|_{s=a_2} \quad ,$$

while A_2 is computed from

$$A_2 = \left. \left| \frac{d}{ds} \frac{P(s)(s-a_2)^2}{Q(s)} \right| \right|_{s=a_2} \quad .$$

44

We see that, in this case, the inverse transform $f(t)$ also contains the term $te^{a_2 t}$. The above can be easily generalized to the case of a root of any arbitrary order $(\leq n)$. This is left as an easy exercise.

APPLICATION TO LINEAR CONSTANT COEFFICIENTS DIFFERENTIAL EQUATIONS

Consider the nth order linear differential equation with constant coefficients

$$\frac{d^n y(t)}{dt^n} + a_{n-1} \frac{d^{n-1} y(t)}{dt^{n-1}} + \cdots + a_0 y(t) = x(t) ,$$

$$t > 0$$

together with the initial conditions:

$$y(0) \quad y'(0), \ldots, y^{(n-1)}(0) ,$$

where, of course, $x(t)$ is given. To solve the equation for $y(t)$ subject to the given initial conditions we can use the Laplace Transform method. Taking the Laplace Transform of the equation we see that we can express the Laplace Transform $Y(s)$ of $y(t)$ in terms of $X(s)$ $(= L_s[x(t)])$ and the initial conditions. Therefore $y(t)$ can be found by taking the inverse transform of $Y(s)$.

Example

Solve

$$\frac{d^2 y(t)}{dt} + y(t) = 1 , \qquad t > 0 ,$$

$$y(0) = 1, \quad y'(0) = 2 \quad .$$

We have, taking the Laplace Transform of both sides of the equation:

$$(s^2+1) \ Y(s) \quad = \quad s + 2 + \frac{1}{s} \quad ,$$

therefore

$$Y(s) \quad = \quad \frac{s + 2}{s^2 + 1} \quad + \quad \frac{1}{s(s^2 + 1)} \quad ,$$

from which y(t) can be found by taking the inverse transform.

ANALYSIS OF LINEAR TIME-INVARIANT AND CAUSAL SYSTEMS BY THE LAPLACE TRANSFORM METHOD

We recall that if h(t) is the impulse response function of a linear time-invariant system, then the system admits the input-output transformation

$$y(t) \quad = \quad \int_{-\infty}^{\infty} h(t-\tau) \ x(\tau) \ d\tau$$

for each t in $(-\infty,\infty)$. Now if the system is also causal, then instead of $h(t-\tau)$ we should have written $h(t-\tau)U(t-\tau)$. Also, since the system is time-invariant, we can take 0 to be the time origin. Therefore an input $x(\cdot)$ can be considered as being applied to the system at time 0, and is taken to be 0 for each t < 0, i.e., $x(\cdot)$ is a "causal" input. Thus for a linear time-invariant and causal system with causal inputs x(t)U(t) we have the input-output transformation:

$$y(t) \; = \; \int_0^t h(t-\tau) \; x(\tau) \; d\tau \; , \qquad \text{for} \quad t \geq 0$$

which, as we have seen, is a convolution integral.

Our main concern here is to obtain a relationship between the Laplace Transform of an input $x(t)$ and that of its corresponding output $y(t)$ of a linear time-invariant and causal system.

THE LAPLACE TRANSFORM OF A CONVOLUTION INTEGRAL AND THE
SYSTEM FUNCTION OF A LINEAR TIME-INVARIANT AND CAUSAL SYSTEM

We now prove the following important theorem.

Theorem

Let $X(s)$, $Y(s)$ and $H(s)$ be the Laplace Transforms of $x(t)$, $y(t)$ and $h(t)$, respectively. If, for each t in $[0, \infty)$:

$$y(t) \; = \; \int_0^t h(t-\tau) \; x(\tau) \; d\tau \; , \qquad (3.12)$$

then

$$Y(s) \; = \; H(s) \; X(s) \quad .$$

Proof.

We have by definition

$$Y(s) \; = \; \int_0^\infty e^{-st} \left\{ \int_0^t h(t-\tau) \; x(\tau) \; d\tau \right\} dt \; ,$$

47

$$= \int_0^\infty e^{-st} \left\{ \int_0^\infty h(t-\tau)\ U(t-\tau)\ x(\tau)\ d\tau \right\} dt \quad .$$

Interchanging the order of integration -- a step that can be justified -- we obtain

$$Y(s)\ =\ \int_0^\infty \left\{ \int_0^\infty e^{-st}\ h(t-\tau)\ U(t-\tau)\ dt \right\} x(\tau)\ d\tau \quad .$$

Setting $z = t-\tau$, the inner integral becomes, for $\tau \geq 0$:

$$e^{-st} \int_0^\infty e^{-sz}\ h(z)\ U(z)\ dz \quad .$$

Therefore

$$Y(s)\ =\ \int_0^\infty e^{-s\tau} \int_0^\infty e^{-sz}\ h(z)\ dz\ x(\tau)\ d\tau \quad ,$$

$$=\ \int_0^\infty e^{-s\tau}\ H(s)\ x(\tau)\ d\tau\ =\ H(s)\ X(s) \quad .$$

This finishes the proof of the theorem.

Note that this Theorem also holds for the case in which $h(t-\tau)$ is a matrix while $x(t)$ and $y(t)$ are vectors. This occurs in multi-input multi-output linear time-invariant systems.

It follows from the above Theorem that a linear time-invariant and causal system is, on the one hand, characterized (in the time-domain) by the impulse response

48

function h(t)U(t),   and on the other hand, <u>character-</u>
<u>ized (in the complex s domain) by the function H(s)</u>
<u>which is just the Laplace Transform of h(t)U(t).</u>   Thus
we define:

<u>Definition</u>

The function H(s) = L_s[h(t)U(t)] is called the
<u>system function</u> of a linear time-invariant and causal sys-
tem.   In particular, if the system is single-input single-
output then

$$H(s) \quad = \quad \frac{Y(s)}{X(s)} \quad , \tag{3.13}$$

where X(s) is the Laplace Transform of an input while
Y(s) is the Laplace Transform of the corresponding output.

The system function H(s) can also be defined as

$$H(s) \quad = \quad \frac{\text{Output of system due to input } e^{st}}{e^{st}} \quad ,$$

$$\text{for} \quad -\infty < t < \infty . \tag{3.14}$$

To see that this is the same as (3.13) we recall that a
linear time-invariant and causal system admits the input-
output transformation

$$y(t) \quad = \quad \int_{-\infty}^{t} h(t-\tau) \; x(\tau) \; d\tau \quad = \quad \int_{0}^{\infty} h(\tau) \; x(t-\tau) \; d\tau \quad ,$$

where, of course, the input x(t) is defined for all t.
Now if x(t) = e^{st}, $-\infty < t < \infty$,   then:

$$y(t) = \int_0^\infty h(\tau) \; e^{s(t-\tau)} \; d\tau \quad ,$$

$$= e^{st} \int_0^\infty e^{-s\tau} h(\tau) \; d\tau \quad ,$$

$$= e^{st} H(s) \quad .$$

Therefore

$$H(s) = L_s[h(t)U(t)] = \frac{y(t) \; (\text{output due to } e^{st})}{e^{st}} \quad ,$$

$$\text{for} \quad -\infty < t < \infty \quad ,$$

as expected.

It is important to note that, for a linear time-invariant system, the input e^{st} results in the output $H(s)e^{st}$ which, except for the multiplicative constant $H(s)$, is the same as the input itself. We will have more to say about this later on.

SYSTEM FUNCTION OF SYSTEMS DESCRIBED BY DIFFERENTIAL EQUATIONS

Let S be a system whose input $x(\cdot)$ and output $y(\cdot)$ are related by the linear constant coefficients differential equation

$$\frac{d^n y}{dt^n} + a_1 \frac{d^{n-1}y}{dt^{n-1}} + \cdots + a_n y(t)$$

$$= \frac{d^m x}{dt^m} + b_1 \frac{d^{m-1}x}{dt^{m-1}} + \cdots + b_m x(t) \quad ,$$

where n and m are arbitrary integers. Such a system
is clearly linear, time-invariant and causal.

To find the system function of S we take x($\cdot$) to
be e^{st}. The equation then becomes

$$\frac{d^n y}{dt^n} + a_1 \frac{d^{n-1} y}{dt^{n-1}} + \cdots + a_n y(t)$$

$$= (s^m + a_1 s^{m-1} + \cdots + b_m) e^{st} \quad .$$

It then follows that y(t) must be of the form
y(t) = Ae^{st}, where A is a constant -- independent of
t -- to be found. Substituting for y in the equation
we obtain

$$A(s^n + a_1 s^{n-1} + \cdots + a_n) e^{st}$$

$$= (s^m + b_1 s^{m-1} + \cdots + b_m) e^{st} \quad .$$

Therefore

$$A = \frac{s^m + b_1 s^{m-1} + \cdots + b_m}{s^n + a_1 s^{n-1} + \cdots + a_n}$$

which is also the system function H(s) of the system.
Of course this can also be found by taking the Laplace
Transform of the differential equation, keeping in mind
that the initial conditions must be zero -- i.e., the
system is at rest.

PROBLEMS

1. Find the Laplace Transforms of

$$f(t) = 6U(t) + 4t\, e^{-t} U(t) \; ;$$

$$f(t) = e^{-3t} \cos (5t+\theta) \; ;$$

$$f(t) = t^2 \sin 4t \; .$$

2. Find the Laplace Transforms of

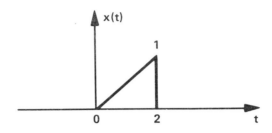

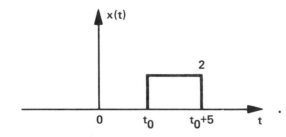

3. Given $f(t) = \sin t$, compute the Laplace Transforms of

$$f(t-t_0) \; ; \qquad\qquad f(t-t_0)\, U(t) \; ;$$

$$f(t)\, U(t-t_0) \; ; \qquad\qquad f(t-t_0) U(t-t_0) \; .$$

4. Let $f(t)$ be a periodic function of period T. Show that

$$F(s) = \frac{1}{1-e^{-sT}} \int_0^T e^{-st}\, f(t)\, dt \; .$$

4. (continued)

Use this result to compute the Laplace Transforms of
the following functions:

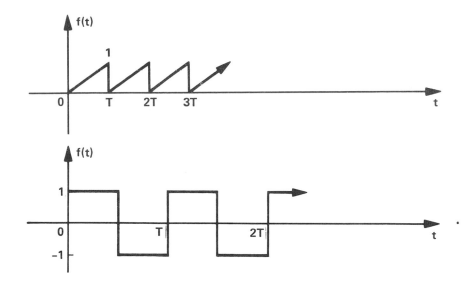

5. Find f(t) given

$$F(s) = \frac{1}{\left(s^2 + a^2\right)^2} , \qquad a: \text{ real constant } ;$$

$$F(s) = \frac{s}{s^4 + 5s^2 + 4} ;$$

$$F(s) = \frac{1}{s^3 + 2s^2 + 10s} .$$

6. Find the inverse Laplace Transforms of the following
functions.

$$\frac{s^2 + 10s + 19}{s^2 + 5s + 6} ; \qquad \frac{3s^2 + 2s + 1}{s^2(s+1)} ;$$

$$\frac{2s}{s^3 + 2s^2 + 1} .$$

7. Solve the following differential equation by the Laplace Transform method

$$\frac{d^2y}{dt^2} + 4y(t) = e^{-t}, \qquad t > 0, \quad y(0) = 0, \quad y'(0) = 1.$$

8. Show that

   (i) $L_S[tf(t)] = \frac{-dF(s)}{ds}$, $\qquad L_S\left[f\left(\frac{t}{a}\right)\right] = aF(as)$,

   $$(a \neq 0).$$

   (ii) For $n \geq 1$:

   $$L_S[(-t)^n f(t)] = \frac{d^n F(s)}{ds^n}.$$

9. Find $f(t)$ given

   $$F(s) = \frac{(1 - e^{-2s})}{s^2 + 5s + 1};$$

   $$F(s) = \frac{1 + 2e^{-s}}{\left(s^2 + 1\right)^2}.$$

10. The figure below is the "poles-zeros" plot of a system function $H(s)$:

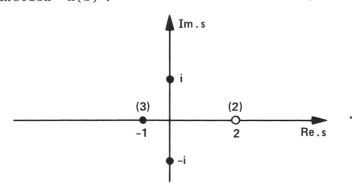

10. (continued)

(i) Find H(s) given that H(0) = $\sqrt{2}$.

(ii) Find the output of the system when tU(t-1) is applied knowing that the system was at rest.

11. The system function H(s) of a linear time-invariant system S is

$$H(s) = \frac{1}{s^2 + s + 1} \ .$$

(i) If x(t) and y(t) are the input and output of S, respectively, find the differential equation relating x(t) and y(t).

(ii) Find y(t) when x(t) = [sin 2(t-1)]U(t-1).

(iii) Find the step response $y_U(t)$ of S -- by direct calculation in the time-domain and then by Laplace Transform.

12. The input x(t) and output y(t) of a linear system S are related by the equation

$$\frac{d^2y(t)}{dt^2} + 3\frac{dy(t)}{dt} + 2y(t) = x(t) , \qquad t > 0 ,$$

$$y(0) = 0 , \qquad y'(0) = 1 \ .$$

(i) Find the impulse response function h(t,σ) of S. Then the response of S when the input U(t-2) was applied -- assuming that the system was at rest.

12. (continued)

   (ii) Solve the differential equation with the given
        initial conditions given $x(t) = te^{-t}U(t)$.

13. In the figure below

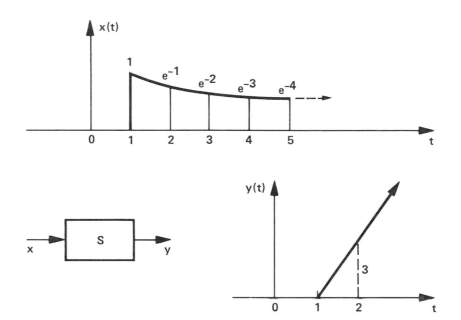

   S is a linear time-invariant system and it was at
   rest when the input shown on the left was applied.
   The corresponding output $y(t)$ is shown on the right.

   (i)   Find the impulse response function $h(t)$ of S.

   (ii) Derive the differential equation which describes
         the system.

   (iii) From the differential equation found in (ii)
         find $y(t)$ when $x(t) = tU(t)$, $y(0) = 0$ and
         $y'(0) = 1$.

14. Find the inverse Laplace Transform of

$$F(s) = \frac{s^2 - (\alpha+\beta)s + \alpha\beta}{s^2 + (\alpha+\beta)s + \alpha\beta} \quad,$$

where α and β are positive constants. Express your answer as a convolution integral. Consider the two cases $\alpha \neq \beta$ and $\alpha = \beta$.

15. Compute

$$y(t) = \int_{-\infty}^{\infty} h(t-\sigma) \, U(t-\sigma) \, x(\sigma) \, d\sigma$$

when

(i) $h(t) = t^2$, $\quad x(t) = e^{-t}U(t-5)$;

(ii) $h(t) = \sin \omega t$, $\quad x(t) = tU(t)$.

16. Compute the equation in problem 15, when

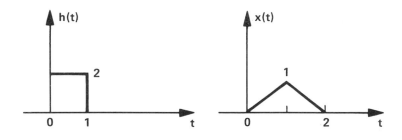

17. Given the input-output relation

$$y(t) = \int_{-\infty}^{\infty} h(t-\sigma) \, x(\sigma) \, d\sigma \quad,$$

where $h(t-\sigma) = \delta(t-\sigma) - e^{-(t-\sigma)}U(t-\sigma)$ and

17. (continued)

$x(t) = te^{-\frac{1}{2}t}U(t).$

   (i)   Find $y(t)$ and its Laplace Transform $Y(s)$.

  (ii)   Compute $H(s) = L_s[h(t)]$ and $L_s[x(t)]$ and verify that the $Y(s)$ found in (i) is equal to $H(s)L_s[x(t)]$.

18. Find the system function and the impulse response function of the system whose input $x(t)$ and output $y(t)$ are related by

$$\frac{d^2y}{dt^2} + 3\frac{dy}{dt} + 2y(t) = 2\frac{dx}{dt} + x(t) \quad .$$

19. Find the output for the system whose system function is $H(s) = \frac{s}{s+1}$ when the input $x(t)$ with

$$L_s[x(t)] = \frac{se^{-2s}}{s^2 + 3}$$

is applied to it.

20. Given $x(t) = \sqrt{2}\, e^{-t}U(t)$ and $y(t) = \sqrt{2}\, e^{-t}U(t)(1-2t)$, find the impulse response function $h(t)$ of the linear time-invariant system which admits the given input-output pair.

21. Using the Laplace Transform, find the convolution

of $f_1(t)$ and $f_2(t)$, where

$$f_1(t) = U(t)e^{-at} \cos(2\omega t - \theta); \quad f_2(t) = U(t)e^{-at},$$

with a and θ positive constants.

22. Determine the <u>output</u> y(t) of the system described by

$$\dot{x}_1(t) = -x_2(t) - 3x_1(t) + u(t) ;$$

$$\dot{x}_2(t) = x_1(t) - x_2(t) - u(t) ,$$

where

$$y(t) = x_1(t) - x_2(t); \quad x_1(0) = 1, \ x_2(0) = -1 ,$$

for the <u>input</u>

$$u(t) = \begin{cases} 1 & 0 \le t \le 2 \\ 0 & \text{elsewhere .} \end{cases}$$

23. (i) Find the Laplace Transform Y(s) of y(t)
given:

$$y(t) = \int_0^t [\delta(t-\tau) - 2e^{-(t-\tau)}U(t-\tau)]\tau^2 \, e^{-\tau} \, d\tau,$$

$$t \ge 0 ;$$

$$y(t) = \int_0^t \sin 2(t-\tau) \, \sigma \cos \sigma \, d\sigma, \quad t \ge 0;$$

$$y(t) = \int_{-\infty}^{\infty} (t-\tau)^2 \cos(t-\tau) \, U(t-\tau) \, e^{-2\tau}U(\tau) \, d\tau .$$

23. (ii) Find f(t) given F(s):

(a) $F(s) = \dfrac{s - 1}{s^2 + 3s + 2}$;

(b) $F(s) = \dfrac{s}{s^2 + s + 2}$;

(c) $F(s) = \dfrac{s^2}{s^2 - 2s + 5}$.

In case (a) express f(t) as a convolution integral.

(iii) Solve the following differential equation by Laplace Transform method:

$$\frac{d^2y(t)}{dt^2} + 5\frac{dy(t)}{dt} + 7y(t) = t , \qquad t \geq 0 ,$$

$y(0) = 0$,   and $y'(0) = 1$.

CHAPTER 4. SIGNALS: FOURIER SERIES ANALYSIS

Analysis of continuous-time signals by the Fourier method is studied in some detail. The Orthogonality Principle which leads to the so-called mean square approximation of a signal by a finite Fourier series will be discussed.

Let $y(\cdot) = T[x(\cdot)]$ be the input-output transformation of a linear system. Then $T[\cdot]$ has the important "additivity" property, i.e., for each $x_1(\cdot)$, $x_2(\cdot)$, and each scalar a_1, a_2: $T[a_1 x_1(\cdot) + a_2 x_2(\cdot)] = a_1 T[x_1(\cdot)] + a_2 T[x_2(\cdot)]$. This, as we know, is the key idea which results in the superposition integral of Chapter 2.

Suppose now that an input $x(\cdot)$ to a linear system can be written as $x(\cdot) = a_1 x_1(\cdot) + a_2 x_2(\cdot)$, where $x_1(\cdot)$ and $x_2(\cdot)$ are, in some suitable sense, "simpler" than $x(\cdot)$. Then to find the response of the system to $x(\cdot)$ we only need to find $T[x_1(\cdot)]$ and $T[x_2(\cdot)]$ which, hopefully, are also simpler than $T[x(\cdot)]$. This leads to the idea of decomposing (or expanding) a signal to a linear system into a linear combination of "elementary" signals, which is the central theme of this Chapter.

ORTHOGONAL DECOMPOSITION OF A SIGNAL

We begin with a simple and familiar situation. Consider the vector $\vec{V} = (V_x, V_y)$ of Figure 4.1. Let $\vec{i}$

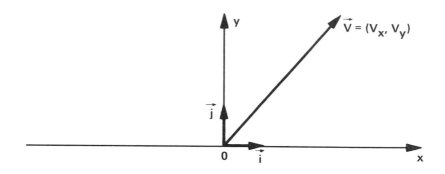

Figure 4.1.

and $\vec{j}$ be the unit vectors along the x and y axes, respectively. Then clearly

$$\vec{V} = V_x \vec{i} + V_y \vec{j} . \tag{4.1}$$

Thus $\vec{V}$ admits a decomposition into a linear combination of the vectors $\vec{i}$ and $\vec{j}$. What is so special about these vectors? They are, by definition, of <u>unit length</u> and <u>perpendicular</u> to each other. These facts can be expressed in terms of the "dot" product as follows

$$|\vec{i}|^2 = \vec{i} \cdot \vec{i} = 1 \tag{4.2a}$$

$$|\vec{j}|^2 = \vec{j} \cdot \vec{j} = 1 \tag{4.2b}$$

and

$$\vec{i} \cdot \vec{j} = \vec{j} \cdot \vec{i} = 0 . \tag{4.3}$$

Two perpendicular vectors are also called <u>orthogonal</u>; in addition, if they are also of unit lengths then they are

62

called orthonormal. We therefore called (4.1) an orthonormal decomposition (or expansion) of the vector $\vec{V}$. It
is important to observe that the scalars V_x and V_y
can be expressed in terms of the vectors $\vec{V}$, $\vec{i}$ and $\vec{j}$.
To see this we only have to take the dot products of (4.1)
with $\vec{i}$ and $\vec{j}$. We find

$$V_x = \vec{V} \cdot \vec{i} \tag{4.4}$$

and

$$V_y = \vec{V} \cdot \vec{j} . \tag{4.5}$$

The above can be generalized to the case of signals.

Let $\{\phi_n(t)\}$, $n = 0, 1, 2, \ldots,$ be a sequence of
complex-valued functions of the time variable t over
[a,b] (say). If for each n:

$$\int_a^b |\phi_n(t)|^2 \, dt = \int_a^b \phi_n(t) \, \overline{\phi_n(t)} \, dt \; < \; \infty$$

where $\overline{}$ denotes the complex conjugate, and for each
$n \neq m$:

$$\int_a^b \phi_n(t) \, \overline{\phi_m(t)} \, dt = 0 .$$

Then ϕ_n is said to be _orthogonal_ to ϕ_m, written
$\phi_n \perp \phi_m$, and $\{\phi_n\}$ is called an _orthogonal_ sequence.
In addition, if

$$\int_a^b |\phi_n(t)|^2 \, dt = 1 , \quad \text{for each} \quad n \geq 0$$

then the sequence is an <u>orthonormal</u> sequence.

Suppose now that we are given a signal $f(t)$ and a sequence of orthonormal signals $\{\phi_n(t)\}$ over the same interval $[a,b]$. If $f(t)$ can be written as

$$f(t) = \sum_{n=0}^{\infty} F_n \, \phi_n(t) \, ,$$

then it is said to admit an <u>orthogonal expansion</u>.

The basic question regarding such an expansion is under what conditions can a signal be so represented? In practice, one of course is interested in "approximating" <u>$f(t)$ by only a finite number of the ϕ_n's.</u> Therefore one has to calculate the "error" involved in such an approximation.

An important orthogonal expansion is the Fourier series representation of periodic signals.

PERIODIC SIGNALS AND FOURIER SERIES

If $f(t)$, $-\infty < t < \infty$, is a signal such that for some fixed $T \ (> 0)$ $f(t) = f(t+T)$, then it is said to be <u>periodic</u> with <u>period</u> T. Clearly for such a signal it is enough to restrict attention to an interval of length T, for instance, $[0,T]$, $\left[-\frac{T}{2}, \frac{T}{2}\right]$ or $[a, a+T]$ for any arbitrary a.

Consider the doubly infinite sequence of complex exponentials
$$\{e^{int}; \quad n = 0, \pm 1, \pm 2, \ldots; \quad -\infty < t < \infty\} \, .$$

64

Then, since for each n: $e^{in(t+2\pi)} = e^{int}$, these functions
are periodic with period 2π. Moreover, for any integers
n, m:

$$\int_{-\pi}^{\pi} e^{int} e^{-imt} \, dt \;\; = \;\; 2\pi \, \frac{\sin (n-m)\pi}{(n-m)\pi} \quad .$$

Therefore

$$\int_{-\pi}^{\pi} e^{int} e^{-imt} \, dt \;\; = \;\; 2\pi \, \delta_{n,m} \;\; = \;\; \begin{cases} 2\pi & \text{for} \quad n = m \\ 0 & \text{for} \quad n \neq m \end{cases} \quad ,$$

where $\delta_{n,m}$ is called the Kronecker delta. We conclude
that the sequence

$$\{e^{int}; \; n = 0, \; \pm1, \; \pm2, \; \ldots; \quad -\pi \leq t \leq \pi\}$$

is orthogonal, equivalently, the sequence $\left\{\dfrac{e^{int}}{\sqrt{2\pi}}\right\}$ is orthonormal.

More generally, it is easy to see that the sequence
of periodic complex exponentials of period T.

$$\left\{e^{in\omega_0 t}; \; n = 0, \; \pm1, \; \pm2, \; \ldots; \quad \omega_0 = \frac{2\pi}{T}, \quad \frac{-T}{2} \leq t \leq \frac{T}{2}\right\} \quad (4.6)$$

is orthogonal

$$\int_{-\frac{T}{2}}^{\frac{T}{2}} e^{in\omega_0 t} e^{-im\omega_0 t} \, dt \;\; = \;\; T \, \delta_{n,m} \quad . \quad (4.7)$$

Let f(t) be a periodic signal with period T and
F_n be scalars given by, for any n:

$$F_n = \frac{1}{T} \int_{-\frac{T}{2}}^{\frac{T}{2}} f(t)\, e^{-in\omega_0 t}\, dt \; . \qquad (4.8)$$

Then the infinite series

$$\sum_{n=-\infty}^{\infty} F_n\, e^{in\omega_0 t} \; , \qquad -\infty < t < \infty \qquad (4.9)$$

is called the <u>Fourier series expansion (or representation)</u> of $f(t)$, while the F_n's are the <u>Fourier coefficients</u> of $f(t)$. Here the function $f(t)$ is, in general, complex-valued.

It is noted that in writing down the infinite series (4.9) we understood that it converges to a periodic function -- of period T, since each term of the series is periodic with the same period. Moreover, there are mild conditions for this function to be the given periodic function $f(t)$. Therefore, in what follows we shall assume that we are only interested in those periodic functions which can be represented by a Fourier series, and we write

$$f(t) = \sum_{n=-\infty}^{\infty} F_n\, e^{in\omega_0 t} \; , \qquad -\infty < t < \infty \; . \qquad (4.10)$$

Note that, since (4.6) is an orthogonal sequence, the Fourier series of $f(t)$ is indeed an orthogonal expansion of $f(t)$.

.

DISCRETE SPECTRA: AMPLITUDE AND PHASE, REAL SIGNALS

The Fourier coefficients F_n of a periodic signal $f(t)$ are generally complex scalars. Therefore we can write, for each n:

$$F_n = |F_n| e^{i\Theta_n} \qquad (4.11)$$

where

$$|F_n|^2 = [Re. \ F_n]^2 + [Im. \ F_n]^2 \qquad (4.12)$$

and

$$\tan \Theta_n = \frac{Im. \ F_n}{Re. \ F_n} . \qquad (4.13)$$

$|F_n|$ is called the amplitude of F_n, while Θ_n is its phase. It is clear that they are functions of the integer n. Hence the plot of $|F_n|$ versus n is called the (discrete) amplitude spectrum of $f(t)$, while that of Θ_n versus n is the phase spectrum of $f(t)$.

Suppose now that $f(t)$ is a real signal, i.e., for each t, $f(t)$ is a real number. Then, since $f(t) = \overline{f(t)}$ it is clear from (4.8) that, for each n:

$$F_{-n} = \overline{F_n} . \qquad (4.14)$$

Therefore, for each n:

$$|F_n| = |F_{-n}| \qquad (4.15)$$

and

$$\Theta_n = -\Theta_{-n} . \qquad (4.16)$$

Thus a real signal must have an even amplitude spectrum and an odd phase spectrum.

It is important to note that each Fourier coefficient F_n is completely determined by its amplitude and phase. Therefore, since $f(t)$ is characterized by its Fourier coefficients, a signal -- which admits a Fourier representation -- is characterized by its amplitude and its phase spectra. In other words, it is possible to "reconstruct" a signal from its spectra.

Now let $f(t)$ be a real periodic signal and we write (4.10) as

$$f(t) = \sum_{n=-\infty}^{-1} F_n \, e^{in\omega_0 t} + \sum_{n=1}^{\infty} F_n \, e^{in\omega_0 t} + F_0 \quad .$$

Therefore

$$f(t) = \sum_{n=1}^{\infty} \left\{ F_{-n} \, e^{-in\omega_0 t} + F_n \, e^{in\omega_0 t} \right\} + F_0 \quad .$$

Substituting for F_{-n} from (4.14) we get

$$f(t) = \sum_{n=1}^{\infty} \left\{ \overline{F}_n \, e^{-in\omega_0 t} + F_n \, e^{in\omega_0 t} \right\} + F_0 \quad .$$

Setting

$$F_n = a_n + ib_n \, , \qquad \text{i.e.,} \quad a_n = \text{Re.} \ F_n \quad \text{and}$$
$$b_n = \text{Im.} \ F_n \quad ,$$

we find

$$f(t) = F_0 + \sum_{n=1}^{\infty} (2a_n) \cos n\omega_0 t + (-2b_n) \sin n\omega_0 t \quad .$$

$$(4.17)$$

The right-hand side of this expression is called a real

Fourier series. The term F_0 -- which is nothing else than the average value of $f(t)$ over a period -- is called the DC term, while

$$(2a_1) \cos \omega_0 t + (-2b_1) \sin \omega_0 t$$

is called the first harmonic -- or the fundamental, and

$$(2a_N) \cos N\omega_0 t + (-2b_N) \sin N\omega_0 t$$

is called the Nth harmonic. The frequency $\underline{\omega_0}$ is called the fundamental frequency.

PROPERTIES

Let $f(t)$ be a real periodic signal, then

$$f(t) = \sum_{-\infty}^{\infty} F_n e^{in\omega_0 t} \quad .$$

We now wish to compute the integral $\frac{1}{T} \int_{-\frac{T}{2}}^{\frac{T}{2}} f^2(t)\, dt$ in terms of the Fourier coefficients F_n's. We have

$$\frac{1}{T} \int_{-\frac{T}{2}}^{\frac{T}{2}} \sum_{-\infty}^{\infty} \sum_{-\infty}^{\infty} F_m F_n e^{i(m+n)\omega_0 t}$$

$$= \sum_{-\infty}^{\infty} F_m \sum_{-\infty}^{\infty} F_n \frac{1}{T} \int_{-\frac{T}{2}}^{\frac{T}{2}} e^{i(m+n)\omega_0 t}\, dt \quad ,$$

$$= \sum_{-\infty}^{\infty} F_n F_{-n} \quad ,$$

where we have made use of the fact that the integral is

equal to T whenever m+n = 0, and is equal to 0 otherwise. But, since f(t) is real, $F_{-n} = \overline{F_n}$, therefore

$$\frac{1}{T} \int_{-\frac{T}{2}}^{\frac{T}{2}} f^2(t)\, dt = \sum_{-\infty}^{\infty} \left| F_n \right|^2 .$$

This relation is known as the Parseval Theorem. The
integral on r.h.s. is called <u>the mean square value</u> of
the real signal f(t). Note that due to this relation we
see that the infinite series on the r.h.s. converges.
Therefore the Fourier coefficients F_n satisfy the
property: $\left| F_n \right|^2 \to 0$, as $n \to \infty$.

MEAN (OR LEAST) SQUARE APPROXIMATION

We now discuss the Orthogonality Principle which is
the key tool of the so-called Mean (or Least) Square
Approximation. This is, perhaps, the most useful method
of approximation in engineering problems.

Let f be a quantity to be "measured"; for instance,
it could be the outcome of an experiment. However, due to
physical or human reasons, the measured values of f --
called the observed data -- are not consistent. Our problem is to find, among the observed data, the "best" estimate for f. This problem has a very simple geometrical
interpretation.

Consider the plane M of Figure 4.2. The point A
is not in M and $\hat{A}$ -- such that $A\hat{A} \perp M$ -- is the only
point of M which is "closest" to A. Thus among all the

70

vectors in M, the vector OÂ can be taken to be the best estimate of the vector OA. Indeed, any vector OB (say) of M can be considered as an estimate of OA, and the error involved is, of course, the length AB. Thus among all these errors, AÂ is the smallest, equivalently, $(A\hat{A})^2$ is the smallest of all the squares of the errors. We therefore say that OÂ is the mean (or least) square estimate (or approximation) of OA and $(A\hat{A})^2$ is the least (or mean) square error.

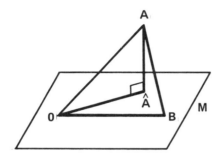

Figure 4.2.

In the above we required that AÂ ⊥ M. This is therefore called the Orthogonality Principle.

Returning to our original problem, the plane M is in this case "generated" by the set of observed data, i.e., M consists of all linear combinations of these data. Thus the problem becomes that of finding the coefficients of a linear combination of the observed data such that the square of the error between f and this linear combination is

71

minimum.

We now "least-square" approximate a periodic function $f(t)$ by a finite sum of the complex exponentials $e^{in\omega_0 t}$. We set

$$f(t) = \sum_{n=-N}^{N} \hat{F}_n e^{in\omega_0 t} \quad ,$$

where the $\hat{F}_n$'s are to be found so that the square of the error between f and $\hat{f}$ is smallest.

Our first task is to decide on an expression for this error. Recall that the orthogonality of $e^{in\omega_0 t}$ and $e^{im\omega_0 t}$, $n \neq m$, is expressed by the integral

$$\int_{-\frac{T}{2}}^{\frac{T}{2}} e^{in\omega_0 t} e^{-im\omega_0 t} dt \quad .$$

Therefore, as in the case of two vectors, the dot (or inner) product of $g(t)$ and $h(t)$ over $\left[-\frac{T}{2}, \frac{T}{2}\right]$ -- written $[g,h]$ -- can be defined to be

$$[g,h] = \int_{-\frac{T}{2}}^{\frac{T}{2}} g(t) \overline{h(t)} dt$$

as long as the integral exists. It then follows -- as in the case of vectors -- the square of the length (or norm) of $f(t)$ -- written $||g(t)||^2$ -- is

$$||g(t)||^2 = \int_{-\frac{T}{2}}^{\frac{T}{2}} |g(t)|^2 dt \quad .$$

The above suggests that we can take the square of the norm of $f - \hat{f}$ to be the square of the error of our approximation. Thus our problem becomes: find the F_n's so that

$$||f - \hat{f}||^2 = \int_{-\frac{T}{2}}^{\frac{T}{2}} |f(t) - \hat{f}(t)|^2 \, dt$$

is smallest.

To apply the Orthogonality Principle we take the plane M to be the family of all linear combinations of the complex exponentials $e^{in\omega_0 t}$, $n = 0, \pm 1, \ldots, \pm N$. Thus we must have

$$f - \hat{f} \perp e^{in\omega_0 t}, \qquad n = 0, \pm 1, \ldots, \pm N .$$

That is, for each n:

$$\int_{-\frac{T}{2}}^{\frac{T}{2}} [f(t) - \hat{f}(t)] e^{-in\omega_0 t} \, dt = 0 .$$

From which is follows that, for $n = 0, \pm 1, \ldots, \pm N$:

$$\hat{F}_n = \frac{1}{T} \int_{-\frac{T}{2}}^{\frac{T}{2}} f(t) e^{-in\omega_0 t} \, dt .$$

Thus $\hat{F}_n$'s are just the Fourier coefficients of the periodic signal $f(t)$. We conclude that each time a periodic signal is approximated by a finite Fourier series expansion, it is actually being least-square approximated.

We note that the coefficients $\hat{F}_n$ can also be found by directly minimizing the quantity

$$\int_{-\frac{T}{2}}^{\frac{T}{2}} |f(t) - \hat{f}(t)|^2 \, dt \quad \left(= \ ||f - \hat{f}||^2 \right) .$$

We have

$$\int_{-\frac{T}{2}}^{\frac{T}{2}} |f(t) - \hat{f}(t)|^2 \, dt = \int_{-\frac{T}{2}}^{\frac{T}{2}} \left| f(t) - \sum_{-N}^{N} \hat{F}_n \, e^{in\omega_0 t} \right|^2 \, dt$$

$$= \int_{-\frac{T}{2}}^{\frac{T}{2}} \left\{ |f(t)|^2 - \sum_{-N}^{N} \overline{\hat{F}_n} \, f(t) \, e^{-in\omega_0 t} - \sum_{-N}^{N} \hat{F}_n \, \overline{f(t)} \, e^{in\omega_0 t} \right\} \, dt$$

$$+ \ T \sum_{-N}^{N} |F_n|^2$$

$$= \int_{-\frac{T}{2}}^{\frac{T}{2}} |f(t)|^2 \, dt \ + \ T \sum_{-N}^{N} |\hat{F}_n|^2 - \overline{\hat{F}_n} F_n - \hat{F}_n \overline{F_n} \quad .$$

Therefore

$$\int_{-\frac{T}{2}}^{\frac{T}{2}} |f(t) - \hat{f}(t)|^2 \, dt$$

$$= \int_{-\frac{T}{2}}^{\frac{T}{2}} |f(t)|^2 \, dt \ + \ T \sum_{-N}^{N} |\hat{F}_n - F_n|^2 \ - \ T \sum_{-N}^{N} |F_n|^2 \quad .$$

The right-hand side is certainly "smallest" when $\hat{F}_n$ is taken to be F_n, as expected.

THE MEAN SQUARE ERROR

The quantity

$$\frac{1}{T} \int_{-\frac{T}{2}}^{\frac{T}{2}} |f(t) - \hat{f}(t)|^2 \, dt$$

is called the mean (or least) square error, and is often denoted by $\overline{\varepsilon_N^2}$. We have from the above:

$$\overline{\varepsilon_N^2} = \frac{1}{T} \int_{-\frac{T}{2}}^{\frac{T}{2}} |f(t)|^2 \, dt - \sum_{-N}^{N} |F_n|^2 .$$

This expression is useful in computing the mean square error. It follows from the Parseval Theorem that

$$\overline{\varepsilon_N^2} = \sum_{-\infty}^{\infty} |F_n|^2 - \sum_{-N}^{N} |F_n|^2$$

$$= \sum_{|n|>N} |F_n|^2$$

which is another expression for the mean square error. It is clear from these results that the mean square error depends on N -- which is the number of terms in the expression of $\hat{f}(t)$. And, of course, the mean square error becomes small as N becomes large.

RESPONSE OF LINEAR TIME-INVARIANT SYSTEMS TO PERIODIC INPUTS

We have shown in Chapter 3 that the response of a linear time-invariant system to the input e^{st}, $-\infty < t < \infty$, is $H(s)e^{st}$, where $H(s)$ is the system function of the system:

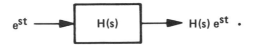

It then follows that if the complex frequency s takes the value $in\omega_0$ then we have

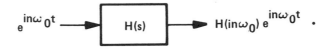

That is, the output of the time-invariant system with system function $H(s)$ -- due to the input $e^{in\omega_0 t}$, $-\infty < t < \infty$ -- is $H(in\omega_0) e^{in\omega_0 t}$. This of course holds for any integer n. We can now prove:

Proposition

A periodic input to a linear time-invariant system always results in a periodic output which has the same period as that of the input.

Proof

Let $x(t)$ be a periodic signal with period T, and let $H(s)$ be the system function of the system. Then we have

$$x(t) = \sum_{-\infty}^{\infty} X_n e^{in\omega_0 t}, \qquad -\infty < t < \infty .$$

Thus it follows from the above discussions and from the fact the system is linear that the corresponding output $y(t)$ is

$$y(t) = \sum_{-\infty}^{\infty} X_n H(in\omega_0) e^{in\omega_0 t}, \quad -\infty < t < \infty,$$

which is clearly periodic with period T. This finishes the proof.

We note that the output $y(t)$ above also admits a Fourier series representation. Set

$$Y_n = X_n H(in\omega_0) = |Y_n| e^{i\Phi_n}$$

(say). Then it is evident that

$$|Y_n| = |X_n| \cdot |H(in\omega_0)|$$

and

$$\Phi_n = \Theta_n + \Psi_n,$$

where Ψ_n is the phase of $H(in\omega_0)$, that is, $H(in\omega_0) = |H(in\omega_0)| e^{i\Psi_n}$. We therefore conclude that the output of a linear time-invariant system -- due to a periodic input -- can be constructed from the amplitude and phase spectra of the input and those of $H(in\omega_0)$.

PROBLEMS

1. Given $f(t) = f(t+T)$ and $f(t) = \sum\limits_{n=-\infty}^{\infty} F_n e^{in\omega_0 t}$,

   show that

   (a) If $f(t) = f\left(t + \dfrac{T}{2}\right)$ then $F_n = 0$ for n odd;

   (b) If $f(t) = -f\left(t + \dfrac{T}{2}\right)$ then $F_n = 0$ for n even.

2. Find and sketch the amplitude and phase spectra of the following signals (only one cycle is shown):

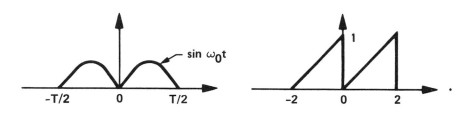

3. Let $f(t) = f(t+T)$ and if

$$f(t) = \sum_{n=-\infty}^{\infty} F_n e^{in\omega_0 t} \quad \text{and} \quad f(t-a) = \sum_{n=-\infty}^{\infty} G_n e^{in\omega_0 t},$$

   where a is a constant.

   Show that $|F_n| = |G_n|$, and that the nth harmonic is shifted in phase by $n\omega_0 a$ radians.

4. Compute the following integrals:

$$\int_{-\pi}^{\pi} \sin kt \cdot \sin \ell t \; dt ;$$

4. (continued)

$$\int_{-\pi}^{\pi} \cos kt \cdot \cos \ell t \quad dt \; ;$$

$$\int_{-\pi}^{\pi} \sin kt \cdot \cos \ell t \quad dt \; .$$

5. Given the periodic signal shown

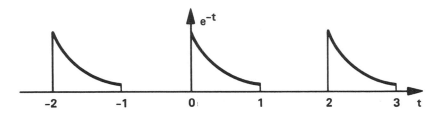

(a) Compute the Fourier coefficient F_n of $f(t)$ and plot the amplitude and phase spectra of $f(t)$.

(b) If $f(t)$ is approximated by the sum $\displaystyle\sum_{n=-2}^{2} F_n e^{in\omega_0 t}$, compute the mean square error $\overline{\varepsilon^2}$.

6. Let $f(t)$ be a periodic function of period T shown below

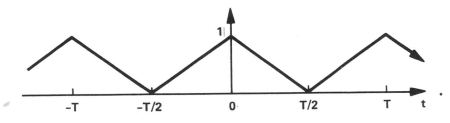

(a) Show that the Fourier coefficients F_n of $f(t)$ are

79

6. (a) (continued)

$$F_n = \begin{cases} \dfrac{2}{n^2\pi^2} & \text{for } n \text{ odd }, \\ 0 & \text{for } n \text{ even and } \neq 0 . \end{cases}$$

(b) Find the mean square error when $f(t)$ is approximated by the first three harmonics, i.e., the finite sum:

$$\sum_{n=-3}^{3} F_n e^{in\omega_0 t} .$$

7. The signal of (6.) is passed through a linear time-invariant system with system function

$$H(s) = \frac{1}{Ts + 1} .$$

Plot the amplitude and phase spectra of the output.

8. (a) Find the Fourier series expansion, the amplitude and phase spectra of the periodic signal shown below.

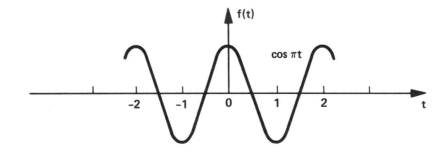

8. (b) Compute the mean square error when $f(t)$ is approximated by

$$F_0 + F_1 e^{i\pi t} + F_{-1} e^{-i\pi t} \quad .$$

(c) The signal above is passed through the linear system whose system function is

$$H(s) = \frac{2s}{2s + 1} \quad .$$

Find the amplitude and phase spectra of the output.

9. Let $f(t)$ be a periodic function with period $T = 2$, and its amplitude and phase spectra as shown below:

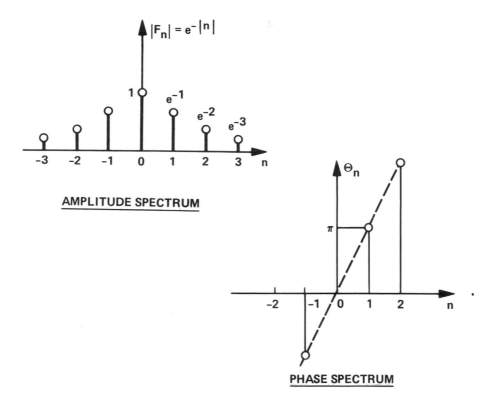

AMPLITUDE SPECTRUM

PHASE SPECTRUM

9. (a) Find the Fourier series representation of $f(t)$,
        and hence $f(t)$.

    (b) Find the mean square error when $f(t)$ is approx-
        imated by

        $$F_1 e^{i\pi t} + F_0 + F_{-1} e^{-i\pi t} .$$

    (c) The function $f(t)$ above is now applied to a
        linear time-invariant system whose impulse re-
        sponse function is

        $$h(t) = \frac{\sin \pi t}{2t} - \frac{\sin \left(\frac{\pi}{2}\right)t}{\pi t} , \qquad -\infty < t < \infty .$$

        Find the corresponding output.

10. (a) Show that the periodic function $f(t)$ below has
        the Fourier series expansion:

        $$f(t) = \frac{\pi^2}{3} + 4 \sum_{n=1}^{\infty} \frac{(-1)^n}{n^2} \cos nt .$$

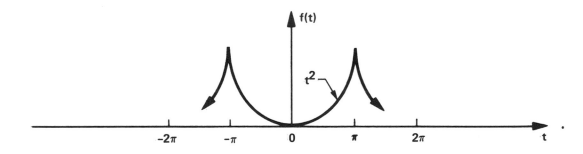

10. (b) Compute the mean square error when $f(t)$ is
approximated by

$$f(t) = \frac{\pi^2}{3} - 4 \cos t + \cos 2t .$$

11. Let $f(t)$ be a periodic function with period T and
let $\omega_0 = \frac{2\pi}{T}$. Find the Fourier coefficients of:

$$\cos \omega_0 t \, f(t) \quad \text{and} \quad \sin \omega_0 t \, f(t)$$

in terms of the Fourier coefficients of $f(t)$.

12. Consider the periodic function $f(t)$ shown below:

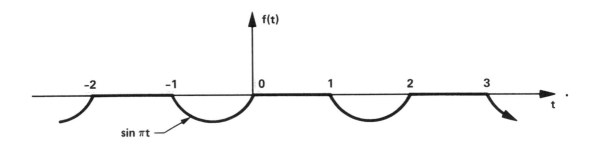

(a) Find the Fourier coefficients F_n of $f(t)$.

(b) The function $f(t)$ is now approximated by

$$g(t) = A_0 + A_1 \cos \pi t + B_1 \sin \pi t .$$

Find A_0, A_1 and B_1 so that $g(t)$ approxi-
mates $f(t)$ in the "mean square sense." Compute the mean square error in this approximation.

13. Show that the periodic function f(t) below:

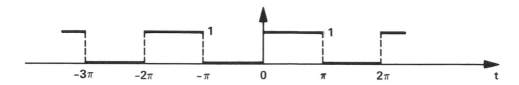

has the Fourier series representation:

$$f(t) = \frac{1}{2} + \frac{2}{\pi} \sum_{k=1}^{\infty} \frac{1}{2k-1} \sin (2k-1)t \quad .$$

This function is now applied to a linear, time-invariant system whose system function is

$$H(s) = \frac{s}{s^2 + 1} \quad .$$

Find the Fourier series representation of the corresponding output.

14. Find the Fourier coefficients F_n of the periodic signal f(t):

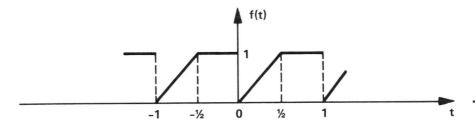

14. (continued)

Find the mean square error when $f(t)$ is approximated by

$$F_0 + F_1 e^{i\pi t} + F_{-1} e^{-i\pi t} \quad .$$

CHAPTER 5. LINEAR TIME-INVARIANT SYSTEMS: FOURIER TRANSFORM ANALYSIS

Analysis of linear time-invariant systems by Fourier Transform is studied. The notion of the frequency (response) function of a linear time-invariant system is introduced. Band-limited signals and the "Sampling" Theorem for this class of signals are also considered.

FOURIER TRANSFORMS

Let $f(t)$, $-\infty < t < \infty$, be a (nonperiodic) signal. The Fourier transform of $f(t)$ denoted by $F\{f(t)\}$ is defined as

$$F\{f(t)\} = \int_{-\infty}^{\infty} e^{-i\omega t} f(t) \, dt \quad (= F(i\omega) \text{, (say))} \quad (5.1)$$

provided of course the integral exists. It is clear that $F(i\omega)$ is a function of the <u>real</u> variable ω, $-\infty < \omega < \infty$ and for each ω, $F(i\omega)$ can be complex.

Example

Let $f(t) = e^{-t}U(t)$, $-\infty < t < \infty$, then

$$F(i\omega) = \int_{-\infty}^{\infty} e^{-i\omega t} e^{-t} U(t) \, dt = \int_{0}^{\infty} e^{-(1+i\omega)t} \, dt$$

$$= \frac{1}{1 + i\omega}$$

from which it follows that

$$\text{Re. } F(i\omega) = \frac{1}{\omega^2 + 1} \quad \text{and} \quad \text{Im. } F(i\omega) = \frac{-\omega}{\omega^2 + 1} .$$

Therefore

$$|F(i\omega)|^2 = \frac{1}{(\omega^2 + 1)^2} + \frac{\omega^2}{(\omega^2 + 1)^2} = \frac{1}{\omega^2 + 1}$$

and

$$\tan \Theta(\omega) = -\omega .$$

Here $|F(i\omega)|$ and $\Theta(\omega)$ are the amplitude and phase of $F(i\omega)$, respectively. They are of course functions of the real variable ω in $(-\infty, \infty)$. Note that $F(i\omega)$ in this example is also equal to the Laplace Transform of $e^{-t}U(t)$ when s is replaced by $i\omega$.

Set $F(i\omega) = |F(i\omega)| e^{i\Theta(\omega)}$, then -- as in the case of the Fourier coefficients F_n -- the amplitude and phase spectra of $f(t)$ are the plots of $|F(i\omega)|$ and $\Theta(\omega)$, they are in this case continuous spectra.

If $F(i\omega)$ is the Fourier Transform of $f(t)$, then $f(t)$ is the inverse (Fourier) Transform of $F(i\omega)$, and is given by

$$f(t) = \frac{1}{2\pi} \int_{-\infty}^{\infty} e^{i\omega t} F(i\omega) \, d\omega . \tag{5.2}$$

Example

Given

$$F(i\omega) = \begin{cases} 1 & \text{for} \quad -\Omega \le \omega \le \Omega \quad , \\ 0 & \text{otherwise} \quad . \end{cases}$$

Then

$$f(t) = \frac{1}{2\pi} \int_{-\Omega}^{\Omega} e^{i\omega t} \, 1 \, d\omega \quad ,$$

$$= \frac{\sin \omega t}{\omega t} \quad , \qquad -\infty < t < \infty \quad .$$

Equation (5.2) is the analog of the Fourier series representation of a periodic function. There, $f(t)$ is represented by an infinite sum of exponential functions $e^{in\omega_0 t}$, while in (5.2) the function $f(t)$ is represented by a continuum of the exponential functions $e^{i\omega t}$. For this reason, $F(i\omega)$ can be considered as the spectrum of the function $f(t)$. Also it is clear that $F(i\omega)$ as defined by equation (5.1) is the continuous analog of the Fourier coefficient F_n.

PROPERTIES

(i) If $f(t)$ is a real-valued function then

$$F(-i\omega) = \overline{F(i\omega)} \quad .$$

(ii) If $F(i\omega)$ is the Fourier Transform of $f(t)$, then the Fourier Transform of $f(t-t_d)$ -- for any t_d -- is

$$F(i\omega) \ e^{-i\omega t_d} \ .$$

(iii) If $F(i\omega)$ is the Fourier Transform of $f(t)$,

then $F(i\omega - i\omega_0)$ is the Fourier Transform of the

function

$$f(t) \ e^{i\omega_0 t} \ .$$

(iv) If $F(i\omega)$ is the Fourier Transform of $f(t)$,

then

$$\text{Fourier Transform of } \frac{d^n f}{dt^n} = (i\omega)^n \ F(i\omega) \ ,$$

for integer $n \geq 0$.

Parseval's Theorem

If $F(i\omega)$ and $G(i\omega)$ are Fourier Transforms of

$f(t)$ and $g(t)$, respectively, then

$$\int_{-\infty}^{\infty} f(t) \ \overline{g(t)} \ dt = \frac{1}{2\pi} \int_{-\infty}^{\infty} F(i\omega) \ \overline{G(i\omega)} \ d\omega \ . \qquad (5.3)$$

Proof

Substituting for $f(t)$ and $g(t)$ in terms of their

transforms -- that is, using (5.2) -- the left-hand side

of (5.3) becomes

$$\int_{-\infty}^{\infty} f(t) \ \overline{g(t)} \ dt$$

$$= \int_{-\infty}^{\infty} \frac{1}{2\pi} \int_{-\infty}^{\infty} e^{i\omega t} \ F(i\omega) \ d\omega \ \frac{1}{2\pi} \int_{-\infty}^{\infty} e^{-i\xi t} \ \overline{G(i\xi)} \ d\xi \ dt \ .$$

Interchanging the order of integration we obtain

$$\int_{-\infty}^{\infty} f(t) \ \overline{g(t)} \ dt$$

$$= \frac{1}{2\pi} \int_{-\infty}^{\infty} \int_{-\infty}^{\infty} F(i\omega) \ \overline{G(i\omega)} \ \frac{1}{2\pi} \int_{-\infty}^{\infty} e^{i(\xi-\omega)t} \ dt \ d\xi \ d\omega \quad .$$

To proceed further we consider the integral

$$\frac{1}{2\pi} \int_{-\infty}^{\infty} e^{i\omega t} \ \delta(\omega-\alpha) \ d\omega \ = \ \frac{1}{2\pi} e^{i\alpha t}$$

which is just the inverse transform of the function $\delta(\omega-\alpha)$. Therefore it follows that

$$\underline{\text{Fourier Transform of}} \ e^{i\alpha t} = 2\pi\delta(\omega-\alpha) \quad . \quad (5.4)$$

Note that with $\alpha = 0$ we get

$$\underline{\text{Fourier Transform of}} \ 1 = 2\pi\delta(\omega) \quad . \quad (5.5)$$

Using (5.4) in our integral above we find

$$\int_{-\infty}^{\infty} f(t) \ \overline{g(t)} \ dt \ = \ \frac{1}{2\pi} \int_{-\infty}^{\infty} \int_{-\infty}^{\infty} F(i\omega) \ \overline{G(i\omega)} \ \delta(\xi-\omega) \ d\xi \ d\omega$$

$$= \ \frac{1}{2\pi} \int_{-\infty}^{\infty} F(i\omega) \ \overline{G(i\omega)} \ d\omega \ .$$

This finishes the proof.

In (5.3) taking $f(t) = g(t)$ we get

$$\int_{-\infty}^{\infty} |f(t)|^2 \, dt \;=\; \frac{1}{2\pi} \int_{-\infty}^{\infty} |F(i\omega)|^2 \, d\omega \quad . \qquad\qquad (5.6)$$

The integral on the left-hand side can be regarded as the
energy contained in the signal $f(t)$. To see this,
we only have to take $f(t)$ to be the voltage across a
one ohm resistor. We conclude from (5.6) that the energy
of a signal is preserved under Fourier Transformation.

ANALYSIS OF LINEAR TIME-INVARIANT SYSTEMS: THE FREQUENCY
(RESPONSE) FUNCTION

Recall that if $H(s)$ is the system function of a
linear time-invariant and causal system then the input
e^{st}, $-\infty < t < \infty$, results in the output $e^{st}H(s)$. It
then follows that, with $s = i\omega$:

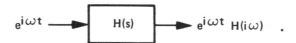

This suggests the following.

Definition

If $H(s)$ is the system function of a linear time-
invariant and causal system then $H(i\omega)$, $-\infty < \omega < \infty$, is
called the <u>frequency (response) function</u> of the system.

Now let $h(t)$, $-\infty < t < \infty$, be the impulse response
function of a linear, time-invariant system -- which need
not be causal. Then, the output due to $e^{i\omega t}$, $-\infty < t < \infty$,

is:

$$\int_{-\infty}^{\infty} h(\tau) \, e^{i\omega(t-\tau)} \, d\tau \;\; = \;\; e^{i\omega t} \int_{-\infty}^{\infty} e^{-i\omega\tau} \, h(\tau) \, d\tau$$

$$= \;\; e^{i\omega t} \{\text{Fourier Transform of } h(t)\}.$$

This suggests that we can generalize the definition of the frequency response function $H(i\omega)$ as in the next definition.

Definition

The Fourier Transform $H(i\omega)$ of the impulse response function $h(t)$ of a linear time-invariant system is called the Frequency Response Function of the system.

This is the exact analog of the relationship between $H(s)$ and the Laplace Transform of $h(t)$ -- for a linear, time-invariant and causal system.

We now prove:

Theorem

Let

$$y(t) \;\; = \;\; \int_{-\infty}^{\infty} h(t-\tau) \, x(\tau) \, d\tau \;\; , \;\; -\infty < t < \infty,$$

and $H(i\omega)$, $X(i\omega)$ and $Y(i\omega)$ be the Fourier Transforms of $h(t)$, $x(t)$ and $y(t)$, respectively. Then

$$Y(i\omega) \;\; = \;\; H(i\omega) \, X(i\omega) \;\; . \tag{5.7}$$

92

Proof

As in the case of the Laplace Transform of a convolution integral, the proof is straightforward. We have

$$Y(i\omega) = \int_{-\infty}^{\infty} e^{-i\omega t} \int_{-\infty}^{\infty} h(t-\tau) \, x(\tau) \, d\tau \, dt$$

$$= \int_{-\infty}^{\infty} x(\tau) \int_{-\infty}^{\infty} e^{-i\omega t} h(t-\tau) \, dt \, d\tau$$

$$= \int_{-\infty}^{\infty} e^{-i\omega\tau} x(\tau) \, d\tau \int_{-\infty}^{\infty} e^{-i\omega\sigma} h(\sigma) \, d\sigma \quad .$$

This finishes the proof of the Theorem.

It follows at once from this Theorem that the Frequency Response Function $H(i\omega)$ of a linear time-invariant system can also be defined as

$$H(i\omega) = \frac{\text{Fourier Transform of output}}{\text{Fourier Transform of (corresponding) input}} \quad .$$

Note that if the system is causal then $h(t) = 0$ for $t < 0$, and we have:

$$H(i\omega) = L_s[h(t)]\Big|_{s=i\omega} = H(s)\Big|_{s=i\omega} \quad ,$$

provided the substitution $s = i\omega$ makes sense (see Notes #2 at the end of the chapter).

To see this we only have to note that ·

$$H(i\omega) \;=\; \int_{-\infty}^{\infty} e^{-i\omega t}\, h(t)\, dt$$

$$=\; \int_{0}^{\infty} e^{-i\omega t}\, h(t)\, dt \quad \text{(by causality)}$$

$$=\; \int_{0}^{\infty} e^{-st}\, h(t)\, dt \bigg|_{s=i\omega} \;=\; H(s)\big|_{s=i\omega} \;.$$

It does not matter whether the system is causal or not; the Frequency Function $H(i\omega)$ can always be found since it is the Fourier Transform of $h(t)$.

Let $\Theta(\omega)$, $\Phi(\omega)$ and $\Psi(\omega)$ be the phases of $X(i\omega)$, $Y(i\omega)$ and $H(i\omega)$, respectively. Then it is evident from the above Theorem that

$$|Y(i\omega)| \;=\; |H(i\omega)| \cdot |X(i\omega)| \tag{5.8}$$

and

$$\Phi(\omega) \;=\; \Psi(\omega) \;+\; \Theta(\omega) \;. \tag{5.9}$$

Equation (5.8) allows one to design a system which, for instance, "passes" all frequencies ω in an interval -- called a band -- and "rejects" all frequencies outside this particular interval. Hence such a system is often called a "filter."

BAND-LIMITED SIGNALS: SAMPLING THEOREM

A signal $f(t)$, $-\infty < t < \infty$, with Fourier Transform $F(i\omega)$ is said to be band-limited if $F(i\omega)$ is identically zero outside a finite interval -- of the ω-axis -- called a band of frequencies. For convenience, we take

94

the "central" frequency to be zero, and therefore

$$F(i\omega) \;=\; 0 \qquad \text{for} \quad |\omega| \geq \Omega \qquad (\text{say}).$$

Such a signal has an amplitude spectrum of the type shown
in Figure 5.1.

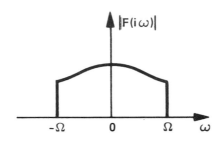

Figure 5.1.

Given a signal f(t) which is <u>not</u> band-limited we
can band-limit it -- to a band of frequencies of length
2Ω -- by passing the signal through an "ideal" low-pass
filter whose frequency function H(iω) is real and is
shown in Figure 5.2.

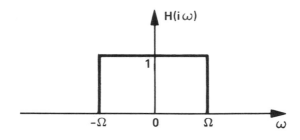

Figure 5.2.

Note that this frequency function has a "zero" phase spectrum, and it "passes" all frequencies Ω in the interval $[-\Omega, \Omega]$, and "rejects" all frequencies outside this interval. Hence the filter is called low-pass. Let $h_\Omega(t)$ be the impulse response functions of the ideal low-pass filter. Then

$$h_\Omega(t) \ = \ \text{Inverse Fourier Transform of} \ \ H(i\omega)$$

$$= \ \frac{1}{2\pi} \int_{-\Omega}^{\Omega} e^{i\omega t} \ 1 \ d\omega$$

$$= \ \frac{\sin \Omega t}{\pi t} \ , \quad -\infty < t < \infty \ .$$

Therefore the output of the low-pass filter due to $f(t)$ is

$$\int_{-\infty}^{\infty} \frac{\sin \Omega(t-\tau)}{\pi(t-\tau)} \ f(\tau) \ d\tau \ , \quad -\infty < t < \infty \ .$$

This signal is denoted by $f_\Omega(t)$ and is called the band-limited version of $f(t)$ -- i.e., it is obtained from the original signal $f(t)$ by a band-limiting process. $f_\Omega(t)$ is, of course, a band-limited signal. It follows from the above discussion that if a signal $g(t)$ is a band-limited signal -- to a band of frequencies 2Ω, then it satisfies the equation

$$\int_{-\infty}^{\infty} \frac{\sin \Omega(t-\tau)}{\pi(t-\tau)} \ g(\tau) \ d\tau \ = \ g(t) \ . \qquad (5.10)$$

Another way of expressing the fact that $g(t)$ is band-limited is, by the inverse transform formula -- (5.2):

$$g(t) = \frac{1}{2\pi} \int_{-\Omega}^{\Omega} e^{i\omega t} G(i\omega) \, d\omega \quad . \tag{5.11}$$

Now let us expand the function $G(i\omega)$ in a Fourier series -- for ω in the interval $[-\Omega,\Omega]$. We have

$$G(i\omega) = \sum_{-\infty}^{\infty} K_n e^{inT_0\omega} \quad , \tag{5.12}$$

where $T_0 = \frac{2\pi}{2\Omega} = \frac{\pi}{\Omega}$, and

$$K_n = \frac{T_0}{2\pi} \int_{-\Omega}^{\Omega} e^{-inT_0\omega} G(i\omega) \, d\omega \quad .$$

It then follows from this and from (5.11) that

$$K_n = T_0 \, g(-nT_0) \quad . \tag{5.13}$$

Substituting (5.13) in (5.12) then (5.12) in (5.11) we find

$$g(t) = \frac{T_0}{2\pi} \sum_{-\infty}^{\infty} g(nT_0) \int_{-\Omega}^{\Omega} e^{i\omega(t-nT_0)} \, d\omega \quad ,$$

or

$$g(t) = \sum_{-\infty}^{\infty} g(nT_0) \frac{\sin \Omega(t-nT_0)}{\Omega(t-nT_0)} \quad . \tag{5.14}$$

We note that in this relation $g(nT_0)$ is just the value of the function $g(t)$ at the time $n\left(\frac{\pi}{\Omega}\right)$ second. Thus we conclude that:

Sampling Theorem

If g(t) is a band-limited signal with a spectrum which is zero above Ω rad/second, then it is determined by its values at the discrete set of points equally spaced at intervals of $T_0 = \frac{\pi}{\Omega}$ second.

PROBLEMS

1. Compute the Fourier Transform of the following functions:

   (a) $f(t) = e^{-a|t|}$, $-\infty < t < \infty$, $a > 0$.

   (b) $g(t) = \begin{cases} e^{i\omega_0 t} , & -T \leq t \leq T \\ 0 , & |t| > T . \end{cases}$

   (c)

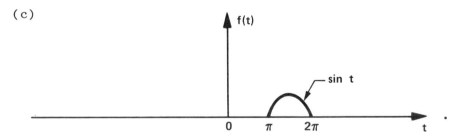

In each case compute and plot the amplitude and phase spectra.

2. If $F(i\omega) = F\{f(t)\}$, show that

$$F\{f(t-a)\} = e^{-i\omega a} F(i\omega) ;$$

$$F\{f(t) \cos \omega_0 t\} = \tfrac{1}{2}[F(i\omega - i\omega_0) + F(i\omega + i\omega_0)] .$$

3. Find $F(i\omega)$ given

$$f(t) = \begin{cases} 1 - \dfrac{|t|}{T} , & -T \le t \le T \\ 0 , & |t| > T . \end{cases}$$

4. Let $f(t)$ be a **real** time function. Discuss the nature of $F\{f(t)\}$ when $f(t)$ is an even function, and when $f(t)$ is an odd function. In each case illustrate your answers by a simple example.

5. Consider the circuit shown

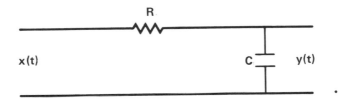

Find $y(t)$ given $x(t) = e^{-|t|}$, $-\infty < t < \infty$.

6. Find the Fourier Transform of the impulse response of a system with output y and input x, governed by the differential equation

6. (continued)

$$6 \frac{d^3y}{dt^3} + 3 \frac{d^2y}{dt^2} + 2 \frac{dy}{dt} + y = 2 \frac{dx}{dt} + x .$$

7. The Fourier Transform $H(i\omega)$ of a function $h(t)$ is shown below.

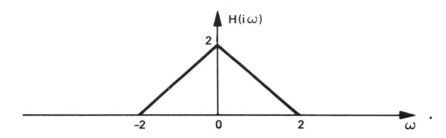

(a) Find $h(t)$.

(b) If $h(t)$ is the impulse response function of a linear time-invariant system, find its output when the input

$$x(t) = \sin (5t + \theta), \quad -\infty < t < \infty ,$$

where θ is a positive constant, is applied.

8. Find the Fourier Transforms of the functions

$$\phi_n(t) = \frac{\sin \Omega(t - nT_0)}{\Omega(t - nT_0)}$$

for $n = 0, \pm 1, \pm 2, \ldots,$ and $-\infty < t < \infty$. Then show

8. (continued)

    that

    $$\int_{-\infty}^{\infty} \phi_n(t) \, \phi_m(t) \, dt \; = \; 0 \qquad \text{for} \quad n \neq m$$

    $$\neq 0 \qquad \text{for} \quad n = m \; .$$

    Therefore given an alternate proof for the sampling
    theorem.

9. Let H(s) be the system function of a linear time-
    invariant system, and let ω_0 be a given fixed fre-
    quency. Find conditions which H(iω) must satisfy
    so that:

    (a) $\underset{-\infty < t < \infty}{\cos \; \omega_0 t}$ ⟶ [H(iω)] ⟶ $\{\cos \; \omega_0 t\}\{H(i\omega_0)\}$.

    (b) $\underset{-\infty < t < \infty}{\sin \; \omega_0 t}$ ⟶ [H(iω)] ⟶ $\{\cos \; \omega_0 t\}\{-iH(i\omega_0)\}$.

    (c) $\underset{-\infty < t < \infty}{\cos \; \omega_0 t}$ ⟶ [H(iω)] ⟶ Re. $\left\{ e^{i\omega_0 t} \, H(i\omega_0) \right\}$.

10. Let h(t), x(t), and y(t) -- for $-\infty < t < \infty$ --
    be the impulse response function, the input and the
    output of a linear time-invariant system, respectively.
    (a) Show that

    $$|Y(i\omega)| \; = \; |H(i\omega)| \; \cdot \; |X(i\omega)|$$

    and

    $$\Theta_Y(\omega) \; = \; \Theta_H(\omega) \; + \; \Theta_X(\omega) \; ,$$

10. (a) (continued)

where H(iω), X(iω) and Y(iω) are the Fourier Transforms of h(t),   x(t)   and   y(t)
respectively, while   ω denotes the phase spectrum.

(b) Given the following "spectra":

INPUT SPECTRA:

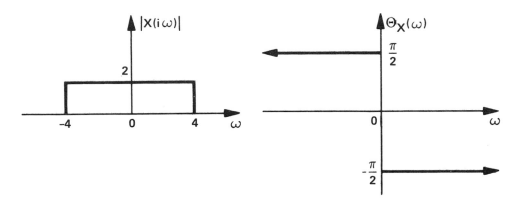

OUTPUT SPECTRA:

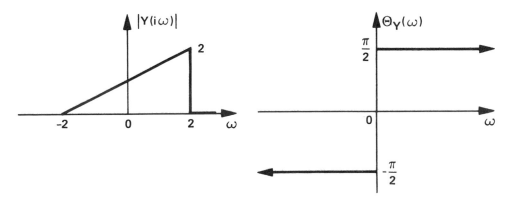

Find H(iω) from the above spectra and from the
fact that H(iω) = 0 for ω __not__ belonging to
the interval (-2,2). Find the impulse response

10. (b) (continued)

function h(t) -- from H(iω) found above.
Is this system causal?

11. Let X(iω) and Y(iω) be the Fourier Transforms of
the input x(t) and output y(t) of a linear time-
invariant system, respectively. The amplitude and
phase spectra of x(t) and y(t) -- i.e., $|X(i\omega)|$,
$\Theta_X(\omega)$, $|Y(i\omega)|$ and $\Theta_Y(\omega)$ -- are shown below:

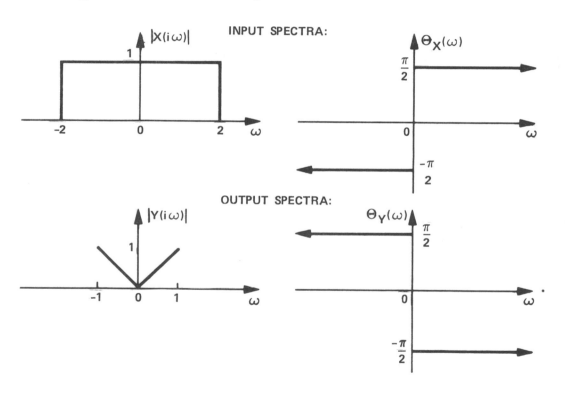

(a) Find the frequency response function H(iω)
of the system. Then find the system impulse

11. (a) (continued)

response function h(t). Is this system causal?

(b) Find the output when the input is the signal:

$(\sin t \; \cos \frac{t}{2})$ for all t.

NOTES

1. Table of Fourier Transforms

f(t)	$F(i\omega) = \mathcal{F}\{f(t)\}$
$\delta(t)$	1
$\text{sgn } t = \dfrac{t}{\|t\|}$	$\dfrac{2}{i\omega}$
$U(t)$	$\dfrac{1}{i\omega} + \pi\delta(\omega)$
1	$2\pi\delta(\omega)$
$f(t) \; e^{i\omega_0 t} \quad (\omega_0 \text{ real})$	$F(\omega - \omega_0)$
$f(t-T) \quad (T \geq 0 \text{ or } < 0)$	$F(\omega) \; e^{-i\omega T}$
$f'(t), \quad f(t) \to 0, \; t \to \pm\infty$	$i\omega F(\omega)$
$\displaystyle\int_{-\infty}^{t} f(\tau) \, d\tau$	$\dfrac{1}{i\omega} F(i\omega) + \pi F(0)\,\delta(\omega)$
$U\left(t+\dfrac{T}{2}\right) - U\left(t-\dfrac{T}{2}\right) \quad (T > 0)$	$\dfrac{T \sin\left(\frac{\omega T}{2}\right)}{\frac{\omega T}{2}}$
$e^{-a\|t\|}, \qquad a > 0$	$\dfrac{2a}{a^2 + \omega^2}$

2. The Fourier Transform exists only under certain conditions on the function, since the defining integral
need not be finite even for simple functions like
$tU(t)$. One sufficient condition is

$$\int_{-\infty}^{\infty} |f(t)| \, dt \; < \; \infty \; .$$

But allowing "generalized" functions like the δ allows one to obtain transforms of functions under weaker
conditions.

This is in contrast to the case of Laplace Transforms which exist for a very large class of functions.
This is so since for the Laplace Transform one can
choose the complex variable s to obtain convergence,
e.g., $\int_{0}^{\infty} t \, dt = \infty$ but $\int_{0}^{\infty} t \, e^{-\sigma t} dt < \infty$ for $\sigma > 0$.
For Fourier Transform, ω is a real variable and
$e^{-i\omega t}$ has absolute value one, so that no choice of
ω helps!

Let $f(t)$ $(= 0$ for $t < 0)$ be such that it has
both a Laplace Transform and a Fourier Transform. What
is the relation between

$$F(s) \; = \; L_s\{f(t)\} \quad \text{and} \quad F(i\omega) \; = \; F\{f(t)\} \; ?$$

Now

$$\hat{F}(i\omega) \; = \; \int_{-\infty}^{\infty} f(t) \, e^{-i\omega t} \, dt \; = \; \int_{0}^{\infty} f(t) \, e^{-i\omega t} \, dt \; ;$$

and

$$F(s) = \int_0^\infty f(t)\, e^{-st}\, dt \;.$$

Comparing the two we obtain

$$\hat{F}(i\omega) = F(s)\big|_{s=i\omega} = F(i\omega)\,,$$

provided we can put $s = i\omega$ in the Laplace Transform, i.e., provided the region of convergence of the Laplace Transform includes points of the form $s = i\omega$, in other words the imaginary axis (Re. $s = 0$).

For Laplace Transforms which are of the rational function (polynomial/polynomial) form, the region of convergence can be decided by looking at the poles: if all the poles are strictly to the left of the imaginary axis then the above relation holds.

Example

(1) $F_\omega\{e^{-at} U(t)\}$, $a > 0$.

$L_s\{e^{-at} U(t)\} = \frac{1}{s+a}$ and if $a > 0$ the transform has a pole at $s = -a$ which is strictly to the left of the imaginary axis in the s-plane. Thus its region of convergence includes the imaginary axis and the Fourier Transform is obtained by:

106

$$F\{f(t)\} \quad = \quad L_s\{f(t)\}\Big|_{s=i\omega} \quad = \quad \frac{1}{s+a}\Big|_{s=i\omega}$$

$$= \quad \frac{1}{a + i\omega} \quad .$$

(2)   As an example of the case where the relation does not hold:

$$L_s\{U(t)\} \quad = \quad \frac{1}{s} \qquad \text{and} \qquad F\{(u(t)\} \quad = \quad \frac{1}{i\omega} + \pi\delta(\omega) \quad .$$

1.   The system S shown below is linear, time-invariant,
     and was at rest when the input x(·) (shown) was
     applied.   The corresponding output y(·) is observed
     and plotted below:

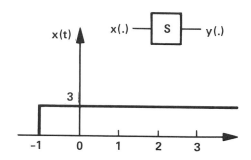

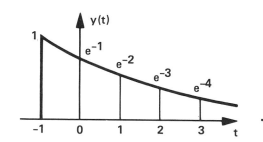

(a)   Find the impulse response function h(t-τ) of S.

(b)   Find the output of S when the input is

$$(t-3) \ e^{-(t-3)} \ U(t-3) \quad .$$

(c)   The system S above is now "cascaded" with a
      second system S_1 (say) as shown:

1. (c) (continued)

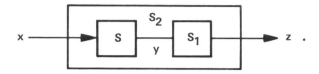

Given that

$$z(t) = \int_0^t t\, y(\sigma)\, d\sigma \quad , \qquad t \geq 0 \quad .$$

Write the relationship between $z(\cdot)$ and $x(\cdot)$, and therefore conclude whether the resulting system S_2 is time-invariant or time-varying. What is the impulse response function $h_2(t,\tau)$ of S_2 ?

2. The input $x(\cdot)$ and corresponding output $y(\cdot)$ of a system S are related via the following equation:

$$\frac{d^2 y(t)}{dt^2} + 4\frac{dy(t)}{dt} + 8y(t)$$

$$= \int_{-\infty}^{\infty} x(t-\tau)\, [\delta(\tau) + 2\tau e^{+\tau} U(t) - 3e^{+\tau} U(\tau)]\, d\tau$$

for $t > 0$, with $y(0) = 0 = y'(0)$ and $x(t) = 0$ for $t < 0$.

(a) Find the relation between the Laplace Transforms of the input and output. Hence, show that the system is linear, time-invariant and causal.

2. (b) What is the system function of the system? Find its poles and zeros (together with their order).

(c) If the input to the system is $x(t) = (\sin t)U(t)$, write down the form of the output $y(t)$ in terms of the undetermined coefficients of the partial fraction expansion. Calculate any two of these coefficients.

3. (a) Show whether the following systems are linear or nonlinear: ($y(\cdot)$ denotes the output of the system and $x(\cdot)$ the corresponding input)

(i) $\dfrac{d^2y(t)}{dt^2} + (\sin t)y(t) = e^{-t}x(t)$,

$$t \geq 0$$

$$y(0) = y'(0) = 0, \quad x(t) = 0, \quad y(t) = 0,$$

$$t < 0.$$

(ii) $y(t) + \displaystyle\int_{-\infty}^{t} \sin^2[y(\tau)]\,\delta(3\tau)\,d\tau + \dfrac{dy(t)}{dt}$

$$= \int_{-\infty}^{t} (t-\sigma)\,x(\sigma)\,d\sigma ;$$

$$y(t) = 0, \quad t \leq 0, \quad x(t) = 0, \quad t < 0.$$

(iii) $y(t) = z(t) + w(t)$, where for $t > 0$

$$\dfrac{dw(t)}{dt} + z(t) = x(t)$$

and

$$\dfrac{dz(t)}{dt} + w(t) = 5x(t)$$

3. (a) (iii) (continued)

and for $t \leq 0$:

$$w(t) = z(t) = x(t) = y(t) = 0 \quad .$$

(b) Is the following system, with input $x(\cdot)$ and corresponding output $y(\cdot)$, physically realizable?

$$\left. \begin{array}{l} \dfrac{dy(t)}{dt} + y(t) = w(t+2) \\[2ex] \dfrac{dw(t)}{dt} - 2w(t) = x(t-2) \end{array} \right\} \quad t > 0$$

$$x(t) = w(t) = y(t) = 0, \quad t \leq 0 \quad .$$

4. Consider the linear system with input $x(\cdot)$ and output $y(\cdot)$ given by the following equation

$$\dfrac{dy(t)}{dt} + 3y(t) = x(t+2) , \quad t > 0$$

with $y(0) = 0$.

(a) Find the input–output relation for the system and hence the impulse response function for the system. We are concerned with the output for $t \geq 0$ only.

(b) Is the system time-invariant? If an extra condition is put on the input: all inputs = 0 for $t < 2$, does your answer change?

(c) Is the system causal?

5. For a given linear system, input-output pairs are
    known as shown below:

$(t - T) U(t -T)$ ⟶ [S] ⟶ $(t^2 - T^2) U(t - T)$

for every T.

(a) Find the impulse response function for the system.

(b) Is S time-invariant?

(c) Is S causal?

(d) What is the output when the input is

$$[\sin t] U(t+3) \ ?$$

6. Given a linear time-invariant causal system S with
    the following information:

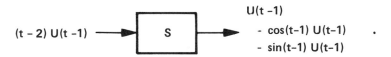

$(t - 2) U(t -1)$ ⟶ [S] ⟶ $\begin{array}{l} U(t -1) \\ - \cos(t-1) U(t-1) \\ - \sin(t-1) U(t-1) \end{array}$.

(a) Find the impulse response $h(t)$ for S, using
     time-domain methods only.

(b) What is the output, when the input is $tU(t)$?
     (Use only time-domain methods.)

(c) Find the Laplace Transform of the following func-
     tion from first principles:

$$[t^2 \cos 2t] \ U(t) \ .$$

6. (d) Find the inverse Laplace Transform of the following functions:

$$\frac{2s + 1}{(s+1)(s^2 + 5s + 2)} \quad ;$$

$$\frac{2s^2 + s - 1}{(s^2 - 2s + 1)s^2} \quad ; \qquad \frac{4s}{(s^2 + 2)^2} \quad .$$

7. Given a linear time-invariant causal system which
yields the output

$$\left\{ e^{-t} - e^{-\frac{1}{2}t} \cos \frac{\sqrt{3}}{2}t + \sqrt{3} \, e^{-\frac{1}{2}t} \sin \frac{\sqrt{3}}{2}t \right\} U(t)$$

when the input is $(1+t)e^{-t}U(t)$, find its transfer
function and impulse response.

If the output is $\delta(t) - e^{-t}U(t)$, what is
corresponding input?

8. Four linear systems S_1, S_2, S_3 and S_4 are connected as shown below:

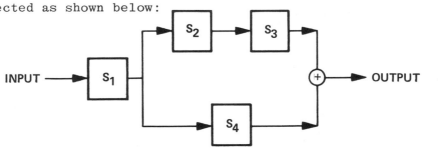

Find the overall impulse response function for the
system given the following descriptions:

S_1: the input $x(\cdot)$ and output $y(\cdot)$ are related
by

113

8. (continued)

S$_1$: (continued)

$$\frac{d^2y(t)}{dt^2} + y(t) = \frac{d^2x(t)}{dt^2} , \qquad t > 0$$

$$y(t) = x(t) = 0 \quad \text{for} \quad t < 0$$

and

$$y(0) = y'(0) = x(0) = x'(0) = 0 \quad .$$

S$_2$: the output is $\frac{1}{t+1}$ times the input for $t \geq 0$,
inputs and outputs are zero for $t < 0$.

S$_3$: the output $y(\cdot)$ corresponding to the input
$x(\cdot)$ is given by the following relation

$$y(t) = \int_0^{3t} (3t+1-\tau) \, x\left(\frac{\tau}{2}\right) \, \delta(2t-\tau) \, d\tau \quad .$$

S$_4$: this is a time-invariant system whose impulse
response is $tU(t)$.

The systems are to be considered for $t \geq 0$.

9. Find the Fourier coefficients of the following periodic
functions:

(a) $f(t) = |\cos 2t|$.

(b)
$$g(t) = \begin{cases} -t+1 , & 0 \leq t \leq 1 \\ 0 , & 1 \leq t \leq 3 \\ t-1 , & 3 \leq t \leq 4 \end{cases} \quad \text{period} = 4 .$$

Plot the amplitude and phase spectra for $f(t)$ and $g(t)$.

10. The function g(t) of problem 9(b) is applied to a linear time-invariant causal system with system function

$$H(s) = \frac{s + 1}{s^2 + s + 1} .$$

Find the Fourier coefficients of the corresponding output.

11. (a) Find the Fourier Coefficients of the "half-wave rectified" sine wave shown below.

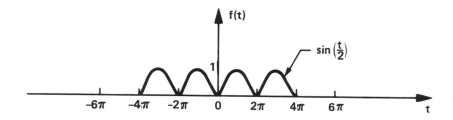

(b) Find the values of the amplitude and the phase at n = ±1.

12. (a) Find the Fourier Coefficients of the periodic function

$$\sum_{n=-\infty}^{\infty} \delta(t-2n-1) = g(t) .$$

(b) The signal g(t) is applied to a linear time-invariant causal system whose system function is $\frac{1}{1+20s}$. Find the Fourier Coefficients of the output.

12. (c) Write down an expression for the mean square error in the output if only the 0th harmonic is retained as an approximation. Using a reasonable approximation, show that this is less than 2.5×10^{-4}.

Hint: $\displaystyle\sum_{n=1}^{\infty} \frac{1}{n^2} = \frac{\pi^2}{6}$.

13. (a) Find the Fourier Transform of $f(t) \sin \omega_0 t$ in terms of the Fourier Transform of $f(t)$.

(b) Show that the Fourier Transform of

$$\int_{-\infty}^{t} f(\sigma)\, d\sigma \quad \text{is} \quad \frac{F(i\omega)}{i\omega} + \pi F(0)\delta(\omega) \;,$$

where $F(i\omega)$ is the Fourier Transform of $f(t)$.

(c) Using (b) and repeated differentiation find the Fourier Transform of

$$\left(1 - \frac{|t|}{T}\right) U(t+T)\, U(T-t) \quad .$$

14. (a) Show that the Laplace Transform of $f(t-T)$ is $e^{-sT}F(s)$ where $F(s)$ is the Laplace Transform of $f(t)$ and $T > 0$.

(b) A linear time-invariant and causal system S whose output is $(\sin t)U(t)$ when the input is $\frac{1}{2}(1+t)U(t)$. Find its system function and impulse response.

14. (c) Find the input of the system   S of part (b)
          knowing that the corresponding output is:

          $2 \sin (t-5)U(t-5) + 4 \sin (t-3)U(t-3) + e^{-t}U(t)$.

15. (a) The input e^{ibt} $(-\infty < t < \infty)$ is applied to a
          linear time-invariant system with frequency
          function $H(i\omega)$, b is a real constant.   What
          is the Fourier Transform of the output?

     (b) Use the result of (a) to find the Fourier
          Series of the output of the system when the
          input is a periodic function with Fourier
          coefficients $\{x_n\}$ and period T.

     (c) The Frequency Function $H(i\omega)$ of a linear
          time-invariant system is shown below:

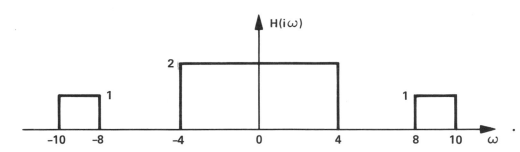

          Find the output when the input is $\sin 6t + \cos 2t$.

16. (a) Find the Laplace Transform of $(\cos t)U(t)$ and
          clearly identify its domain of convergence.

16. (b) Solve the following equation for $y(t)$:

$$\frac{dy(t)}{dt} + \int_0^t y(\sigma) [6U(t-\sigma) + 9\delta(t-\sigma)] \, d\sigma$$

$$= tU(t) \qquad \text{for} \quad t > 0,$$

with $y(0) = 0$, $y(t) = 0$, $t < 0$.

17. (a) Find the Frequency Function for the following linear time-invariant system:

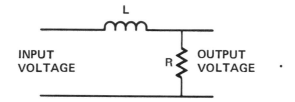

(b) Find the output when the input is $1 + e^{-|t-1|}$ $(-\infty < t < \infty)$ and $L = 1$, $R = 1$.

18. (a) The following input and output are observed for a linear time-invariant system S.

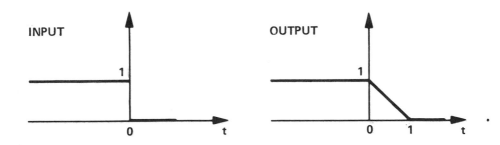

18.   (a)   (continued)

Find the impulse response without using any transform method.   Is the system causal?

(b)   Find the output when the input is $tU(t)$.

(c)   The system   S   is cascaded with a system S_1 as shown below:

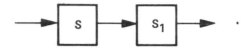

The system S_1 has an impulse response function $h_1(t,\tau) = tU(t-\tau)$.   What is the impulse response of the overall system?

19.   (a)   Show that the Fourier Transform of $f(t-T)$ is $e^{-i\omega T}F(i\omega)$ where $F(i\omega)$ is the Fourier Transform of $f(t)$ and   T   is a real constant.

(b)   The input $e^t U(-t)$ to a linear time-invariant system   S   results in the output

$$U(t) + e^t U(-t) - U(t-1) - e^{t-1}U(1-t) \quad .$$

Find the frequency function for   S   and its impulse response function.

(c)   Is the system   S   causal?   Find its output, when the input is $tU(t)$.

20.   (a)   Find the Fourier Coefficients of the periodic function $f(t)$ shown on the next page:

20. (a) (continued)

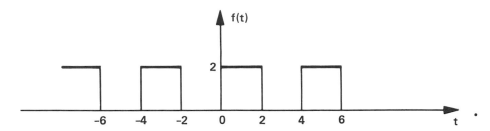

(b) If the Fourier coefficients of a periodic function $x(t)$ are $\{x_n\}$, show that those of the
function $x(t-t_0)$ are

$$\left\{ e^{-int_0\frac{2\pi}{T}} x_n \right\} ,$$

where t_0 is a real constant and T is the
period. Use this and results of part (a) to
find the Fourier coefficients of the function
$g(t)$ sketched below:

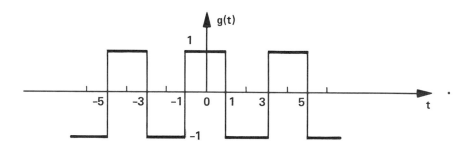

(c) The signal $g(t)$ above is applied to a linear
system with system function $\dfrac{s + 1}{s^2 + 2s + 2}$.
Find the Fourier coefficients of the output.

21. The input to a linear time-invariant system has the amplitude and phase spectra shown below

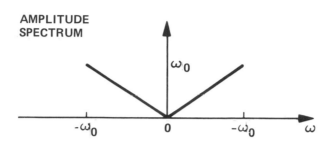

AMPLITUDE SPECTRUM

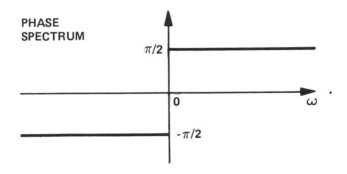

PHASE SPECTRUM

The corresponding output has the amplitude spectrum

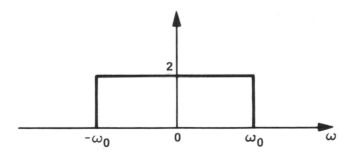

and phase spectrum = 0. This is true for every constant ω_0.

(a) Find the frequency function for the system.

(b) Find its impulse response.

121

21. (c) Find the input when the output is $e^{-|t|}$

$(-\infty < t < \infty)$.

22. Show clearly whether the following systems are
linear, time-invariant, causal, or not. $x(\cdot)$ is
the input and $y(\cdot)$ the corresponding output.

$$S_1: \quad y(t) = \int_{-\infty}^{t} e^{-|t-\tau|} \sin (x(\tau)) \, d\tau$$

$$S_2: \quad y(t) = e^{-t} \int_{-\infty}^{-t} e^{-\sigma} x(-\sigma) \, d\sigma$$

$$S_3: \quad y(t) = \int_{0}^{1} (\sin \tau) \, x(t-\tau) \, d\tau \quad .$$

23. (a) The pole-zero plot of the system function $H(s)$
of a system S is shown below:

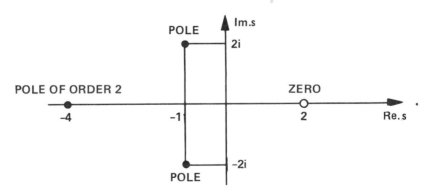

It is known that the average value over a period of the output is $\dfrac{5}{\pi}$ when the input is
$|\cos t|$. Find $H(s)$.

23. (b) Find the impulse response function of S.

(c) What is the output, when the input is identically 1 for all t in $(-\infty, \infty)$?

24. Two systems S_1 and S_2 are described by the following input-output relations: (for $-\infty < t < \infty$)

S_1: $y(t) = x(t) \cos(tx(t))$;

S_2: $y(t) = \displaystyle\int_{-\infty}^{t} \sin(t-\tau) \, e^{-(t-\tau)} \frac{x(\tau)}{1+\tau^2} \delta(\tau) \, d\tau$.

(a) Show whether the systems S_1 and S_2 are linear and causal.

(b) S_1 and S_2 are connected together to form systems S_3 and S_4 as shown below:

Obtain the input-output relations for S_3 and S_4.

(c) Determine whether S_3 and S_4 are linear and causal.

25. A signal $s(t)$ is applied to a linear time-invariant system whose impulse response is $e^{-2|t|}$. The amplitude spectrum of $s(t)$ is shown on the next page:

25. (continued)

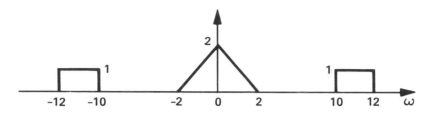

(a) Find and sketch the amplitude spectrum of the output.

(b) Find the energy in the input signal and the percentage energy in the frequency range $|\omega| > 5$.

(c) Write an expression for the energy of the output signal and find the energy in the frequency range $|\omega| > 5$, approximately. (In the second part you may approximate an integral like
$\int_{a}^{b} R(\omega) \, d\omega$ by $R(a)(b-a)$ for $a > 0$, $b > 0$, for example.)

26. Consider the circuit:

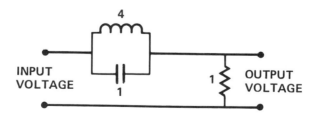

(a) Find its system function.

(b) Find its impulse response.

26. (c) What is the output, when the input is

(sin $\frac{1}{2}$t)U(t)?

27. For the function f(t) shown:

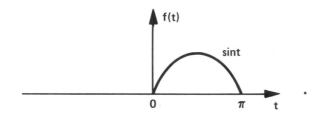

(a) Compute the Laplace Transform of f(t).

(b) Compute $L_s\{f(t)\}$ using the transform of (sin t)U(t) and any theorems concerning Laplace Transforms.

(c) Solve the following equation for x(t):

$$f(t) = \int_{-\infty}^{\infty} f(\tau) \ U(t-\tau) \ x(t-\tau) \ U(\tau) \ d\tau$$

$$- \int_{-\infty}^{t-\pi} \sin \ (t-\pi-\tau) \ x(\tau) \ U(\tau) \ d\tau \ ,$$

for t $\geq$ 0

and x(t) = 0 for t < 0 (f(t) is as in (a)).

28. The input x($\cdot$) and output y($\cdot$) of a linear time-invariant system satisfy:

$$\begin{cases} \dfrac{d^2y}{dt^2} + 3\dfrac{dy}{dt} + 2y(t) = x(t), & t > 0 \\ \\ y(0) = y'(0) = 0. \end{cases}$$

28. (continued)

Find the output when the input is

$e^{2t}U(-t) + (t+1)U(t)$.

29. (a) Find the inverse Fourier Transform of the function $U(\omega+2) - U(\omega-2)$ from the definition.

(b) A linear time-invariant system has an impulse
response function

$$h(t) = \frac{\sin 2t}{\pi t} \cos 6t \qquad (-\infty < t < \infty) .$$

Find its frequency function.

(c) Find the output, when the input is

$$t + \left[\cos\left(\frac{5t}{2}\right)\right]^2 + \cos 2t .$$

30. (a) Prove the orthogonality relation for complex
exponentials:

$$\frac{1}{T} \int_0^T e^{in\omega_0 t} e^{-im\omega_0 t} dt = \delta_{nm} ,$$

where $\omega_0 = \frac{2\pi}{T}$ and $\delta_{nm} = 1$ for $n = m$ and
zero otherwise.

(b) Find the Fourier coefficients of the periodic
signal shown on the next page:

30. (b) (continued)

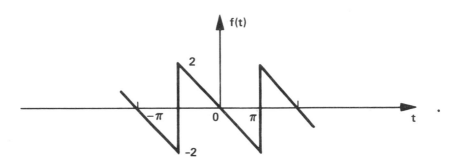

(c) Sketch the amplitude and phase spectra of f(t).

31. The system function of a linear time-invariant system
    has a pole of order 2 at -3 and a pole each at
    -1 + 2i and -1 - 2i. It is also known that the
    average value (over a period) of the output of the
    system is 18 when the input is: 3 + sin 7t + cos 3t.
    (a) Find the system function of the system.
    (b) Find the output of the system when the input
        is $\delta(t) + 2e^t U(t) - 2e^{-3t} U(t)$.
    (c) Find the differential equation relating the
        input and output of the system.

32. (a) Show that if f(t), $-\infty < t < \infty$, is an even
        function then its Fourier Transform is given by:

        $$F(f(t)) = \int_{-\infty}^{\infty} f(t) \cos \omega t \; dt \; .$$

    (b) The system function of a linear time-invariant

32. (b) (continued)

system is $\dfrac{s^2 + 16}{s^2 + 5s + 6}$. The input to this

system is $4 + 3\cos t + \sin 3t + 5 \sin 4t$. Plot

the discrete amplitude spectra of the input

and output.

(c) Find the mean square error in the output if

terms up to the third harmonic are retained in

the Fourier Series, in the context of (b) above.

33. (a) If $h(t,\tau)$ is the impulse response function of

a linear system, show that for a time-invariant

system $h(t,\tau)$ is a function of $t-\tau$.

(b) Obtain the relation $Y(s) = H(s)X(s)$ for a

linear time-invariant causal system from the

time-domain relation.

(c) The input-output relation for a linear system is

$$y(t) = \int_{-2}^{2} x(t-\tau)\,(\tau-1)^2\,d\tau \quad .$$

Is the system time-invariant?

34. The circuit given below is considered to be a system

with the current $I(\cdot)$ as input and the voltage

$V(\cdot)$ as output.

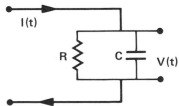

34. (a) Find the input-output relation for this system explicitly.

    (b) Show whether the system is linear or not.

    (c) Clearly show whether the system is causal.

35. When the input $U(t-T)$ is applied to a given linear system, the corresponding output is $\frac{3t+T}{2}(t-T) U(t-T)$ for every real T.

    (a) Find the impulse response function for this system.

    (b) Using the impulse response or otherwise, explain whether the system is

        (i) causal or not ;

        (ii) time-invariant or time-varying.

36. It is known that the following input-output pairs correspond to a given system:

input $x(\cdot)$	corresponding output $y(\cdot)$
$[\sin t]U(t)$	$[\cos t]U(t)$
$[t + \sin t]U(t)$	$[3 \cos t]U(t)$
$tU(t)$	$2tU(t)$
$(t-5) U(t-5)$	$(2t+1) U(t-1)$

    (a) Is the system linear?

    (b) Is the system causal?

    (c) Is the system time-invariant?

    Explain your answers.

129

37. Find the Laplace Transform of the following functions:

(a) $[\cos t]U(t)$.

(b) $\delta(t-T)$ where $T > 0$ is fixed.

38. For a given system, the input $x(\cdot)$ and the corresponding output $y(\cdot)$ are related by the following equation.

$$2 \frac{d^2 y(t)}{dt^2} + 10\frac{dy(t)}{dt} + 8y(t) = \int_0^t e^{-(t-\tau)} x(\tau) \, d\tau$$

for $t \geq 0$ with $y'(0) = 0$ and $y(0) = 0$. All inputs and outputs are zero before time zero.

(a) Find the relation between $Y(s) \equiv L_s\{y(t)\}$ and $X(s) = L_s\{x(t)\}$.

(b) Find the output $y(t)$ of the system, when the input is $[e^{-t} \cos t]U(t)$.

(c) Find the impulse response $h(t)$ of the system.

39. (a) Show that $L_s\{D_{-T}f(t)\} = e^{sT} L_s\{f(t)\}$ for every $T \geq 0$ provided $f(t) = 0$ for $t < T$, here $D_{-T}f(t) = f(t+T)$.

(b) The input $w(\cdot)$ and the corresponding output $v(\cdot)$ for a system are related by the following differential equation

$$\frac{dv(t)}{dt} + 2v(t) = w(t+1) , \qquad t \geq 0 .$$

All inputs $w(t)$ are zero for $t < 1$, and

39. (b) (continued)

outputs $v(t)$ are zero for $t < 0$, and $v(0) = 0$. Find the relation between $V(s) = L_S\{v(t)\}$ and $W(s) = L_S\{w(t)\}$.

(c) Express the relation obtained in (b) in the time-domain, i.e., find the explicit relation between $w(t)$ and $v(t)$.

(d) Assuming that the relation obtained in (c) is valid for inputs $w(t)$ which are zero for $t < 0$ (and may be nonzero for $0 \geq t \geq 1$), i.e., assuming that the differential equation in (b) holds for such inputs, find the impulse response $h(t)$ of the system.

(e) Why can't the impulse response be obtained for the system using the relation obtained in (b) by the usual method of putting $W(s) = 1$?

40. (a) Show that $L_S\left\{\dfrac{df(t)}{dt}\right\} = s L_S\{f(t)\} - f(0)$ for a function $f(t)$ satisfying $|f(t)| \leq Me^{ct}$ for some finite M and c.

(b) Solve the following equation for $g(t)$:

$$f(t) = \frac{dg(t)}{dt}$$

$$+ \int_{-\infty}^{\infty} U(t-\tau+2)\ \sin 2(t-\tau+2)\ g(\tau-2)$$

$$\times\ U(\tau-2)\ d\tau\ ,$$

where $g(0) = 0$, $f(t) = \delta(t) + \tfrac{1}{2}[1 - \cos 2t]U(t)$.

131

40. (b) (continued)

(Attempt partial fraction expansion only after
you have simplified the expression for G(s)
as much as possible.)

41. (a) Find the Fourier coefficients for the signal
f(t) shown below:

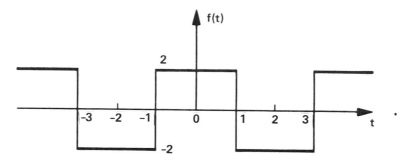

(b) The signal f(t) of (a) is applied to a linear
system S whose description is given below.
Find the Fourier coefficients of the output.

x(t) ———▶ S ———▶ y(t)

$$\begin{cases} y(t) + \int_0^t y(\tau)\ d\tau = x(t) , & t \geq 0 \\ x(t) = 0 = y(t) , & t < 0 . \end{cases}$$

42. A linear time-invariant causal system admits an output
$\frac{1}{9}(1 - \cos 3t)U(t)$, when the input is $tU(t)$.

(a) Find the system function of the system.

42. (b) What is the output when the input is U(t)?

43. For a given linear time-invariant system, the input
    $e^{-t}U(t)$ results in an output

    $$\left(1 - e^{-(t+T)}\right)U(t+T) - \left(1 - e^{-(t-T)}\right)U(t-T) \quad,$$

    where T > 0 is some fixed number.

    (a) Find the impulse response of the system.

    (b) What is the output of the system when the input
        is $e^{-|t|}$?

    (c) What can you say about the causality of the
        system?

44. A given time-invariant linear system has as output
    the function

    $$\frac{2}{\sqrt{3}} e^{-\frac{1}{2}t}\left[\sin \frac{\sqrt{3}}{2}t\right] U(t) \quad,$$

    when the input is $10\delta(t) + U(t)$.

    (a) Find the frequency function for the system.

    (b) When the input is $1 + \cos t + 3\sin 5t$, what
        is the mean square error in the output if only
        terms up to the third harmonic are retained
        in the Fourier series of the output?

    (c) Is the system causal?

45. A linear time-invariant system has the following
    input-output pair:

45. (continued)

$$(\tfrac{1}{2} \operatorname{sgn} t + e^t U(-t), \quad \tfrac{1}{2} e^{-|t|}) \quad .$$

(a)   Find the impulse response for this system.

(b)   What is the output, when the input is   tU(t)?

(c)   Is the system physically realizable?

46.   An idealized version of an amplitude modulation

   (A.M.) communication system is shown below:

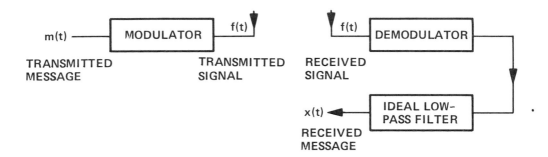

The modulator and demodulator are linear systems

with input-output relations as shown below:

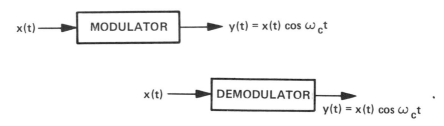

The transmitted and received signals are identical

(f(t))   and the amplitude spectrum of the message

m(t)   is shown on the next page:

46. (continued)

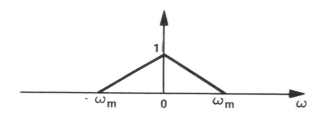

where ω_m and ω_c are constant frequencies with
$\omega_c \gg \omega_m$. (For the purposes of this problem, you
may take ω_c to be 3 or 4 times ω_m.)

The low-pass filter is a linear time-invariant
system with frequency function $L(i\omega)$ whose amplitude spectrum is sketched below:

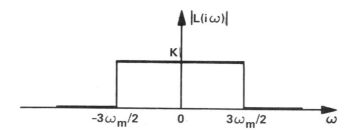

where K > 0 is a finite constant.

Sketch the amplitude spectra for all the signals
involved: m(t), f(t), g(t), and r(t); hence,
show that the amplitude spectrum of the received
message is the same as that of the transmitted message except for a factor of K.

47. The signal $3 + \cos 5t + \sin 10t + \sin t$ is applied to a linear time-invariant system whose impulse response is shown below:

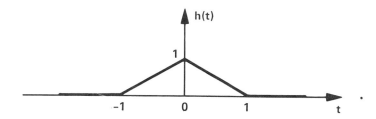

(a) Calculate the frequency function for the system.

(b) What is the average value of the output?

(c) Find the mean square error in the output when only terms up to the seventh harmonic are retained in the Fourier Series.

48. Two systems, S_1 and S_2, are known to be of the form

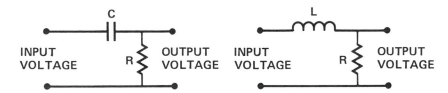

but it is not known which is which. The signal $|\sin \pi t|$ is applied as an input to both systems S_1 and S_2 and it is found that the 0th, first and second harmonic terms of the amplitude spectra of the outputs are

$$0, \quad \frac{\sqrt{2}}{3\pi}, \quad \frac{4}{3\sqrt{5}\pi}, \quad \text{for} \quad S_1$$

and

48. (continued)

$$\frac{2}{\pi} \quad , \quad \frac{\sqrt{2}}{3\pi} \quad , \quad \frac{2}{3\pi\sqrt{5}} \quad , \quad \text{for } S_2 \quad .$$

Find the individual forms of the systems S_1 and S_2.

49. Given the first order differential equation

$$\begin{cases} \dfrac{dy(t)}{dt} + \alpha(t)y(t) = x(t) , & t > 0 \\[2mm] y(0) = 0 \end{cases}$$

where $\alpha(t)$ is a given function.

(a) Express $y(t)$ in terms of $x(t)$ and $\alpha(t)$.

(b) Using the result found in (a), find $y(t)$ when $x(t) = \delta(t-\tau)$, where $t \geq \tau$, and $-\infty < t, \quad \tau < +\infty$.

50. (a) Sketch the following functions for $-\infty < t < \infty$:

$$f(t) = -4\delta(t+5) + e^{-t}U(t-2) ;$$

$$g(t) = e^{-2t} tU(t-4) U(6-t) ;$$

$$h(t) = [\sin 2t]\delta(t) + [\cos 3t]U(4-t) .$$

(b) Compute the following integrals:

$$I_1(t,\tau) = \int_{-\infty}^{\infty} e^{-(t-\sigma)} U(t-\sigma) \sigma U(\sigma-\tau) d\sigma$$

for $-\infty < t, \sigma, \tau < \infty$;

50. (continued)

$$I_2(t) = \int_{-\infty}^{\infty} \sin(t-\tau) \, U(t-\tau) \, e^{-2\tau} \, U(\tau) \, d\tau$$

for $-\infty < t, \quad \tau < \infty.$

51. Write an expression for each of the following functions:

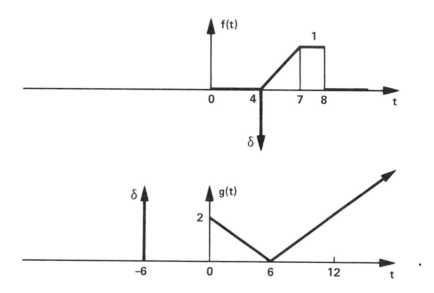

52. Verify whether the following systems are: linear, time-varying, and memoryless. $y(\cdot) = T[x(\cdot)]$.

(a) $y(t) = \int_{-\infty}^{t} t\tau x(\tau) \, U(\tau-t) \, d\tau$.

52. (b) $y(t) = e^{x(t)}$, $-\infty < t < \infty$.

(c) $y(t) = \int_{-\infty}^{\infty} (t-\sigma)^2 \; U(t-\sigma) \; x(\sigma) \; U(\sigma) \; d\sigma$.

53. For the circuit shown below, $x(\cdot)$ is considered as the input and $y(\cdot)$ the output:

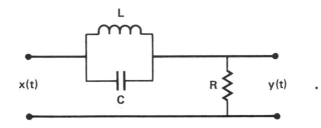

(a) Find the system function of the system. What are its poles and zeros?

(b) Find the impulse response of the system. The cases $\frac{L}{R} = 4RC$ and $\frac{L}{R} \neq 4RC$ will have to be considered separately.

54. Let $f(t)$ be the periodic function sketched below:

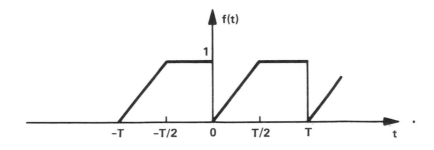

Find the Fourier coefficients for $f(t)$ and sketch

54. (continued)

the phase and amplitude spectra.

55. (a) A linear time-invariant causal system S is
         described by the following equation between
         its input x(t) and the corresponding output
         y(t), when they satisfy the condition
         x(t) = 0 = y(t), t < 0;

$$\frac{dy(t)}{dt} + \int_0^t y(t-\tau) \ e^{-\tau} \ d\tau \ = \ x(t) \ ,$$

                                    for t > 0 and y(0) = 0 .

    Find the system function for the system.

    (b) The periodic signal $|\sin t| + 1$ ($-\infty < t < \infty$)
         is given as an input to the system S of
         part (a). Find the Fourier coefficients of
         the output.

56. (a) A periodic function f(t), with period T,
         has Fourier coefficients $\{f_n, \ n = 0, \pm 1, \pm 2, \ldots\}$.
         Find the value of the following integral in
         terms of these coefficients:

$$\int_3^{3+2T} f(t) \left[\cos \frac{10\pi}{T}t\right] \ dt \ .$$

    (b) Show that Fourier coefficients of a real even
         periodic function are real, and those of a real

140

56. (b) (continued)

odd periodic function are imaginary.

57. Let g(t) be the periodic function

$$g(t) = \sum_{n=-\infty}^{\infty} \left\{ \frac{t-2nT}{T}[U(t-2nT) - U(t-(2n+1)T)] \right.$$

$$\left. + \left[2 - \frac{t-2nT}{T}\right]U(t - (2n+1)T)U((2n+1)T - t)\right\},$$

where T > 0 is a constant.

(a) Sketch the function g(t). What is its period?

(b) Find the Fourier coefficients of g(t).

58. A linear time-invariant causal system produces an
output of $e^{-2t}(\sin t)U(t)$ when the input is
$\delta(t) + \frac{1}{2}U(t)$.

(a) Find the transfer function H(s) of the system.

(b) The input: $2 + \cos 2t + 6 \sin 6t$ is applied to
the system. What is the average value (over a
period) of the output? What is the mean square
error in the output if it is approximated by
terms up to the third harmonic?

59. (a) Find the Laplace Transform of the following
functions:

$$(t^2 \sin t)U(t) ,$$

$$te^{-2t} U(t) U(t-2) .$$

59. (b) Using the convolution integral, find $f(t)$ when its Laplace Transform $F(s)$ is

$$F(s) = \frac{1}{s^2(s+2)} .$$

Verify your result by taking the inverse transform of $F(s)$ directly.

60. For the following system:

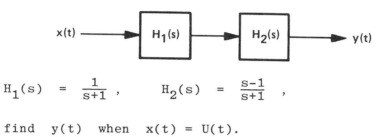

$$H_1(s) = \frac{1}{s+1} , \qquad H_2(s) = \frac{s-1}{s+1} ,$$

find $y(t)$ when $x(t) = U(t)$.

61. Consider the following system:

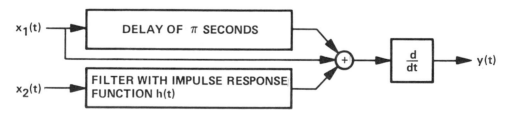

where the function h is given by

$$h(t) = \begin{cases} 0 , & t < 0 \\ 1 , & 0 < t < 10 \\ 0 , & t > 10 \end{cases} .$$

Obtain and sketch the output waveform $y(t)$ when the two inputs are given by

142

61. (continued)

$$x_1(t) = \begin{cases} 0, & t < 0 \\ \sin t, & 0 < t < 2\pi \\ 0, & t > 2\pi \end{cases}$$

$$x_2(t) = U(t).$$

62. Consider the following three functions:

$$x_1(t) = e^{i2\pi t} \quad ;$$

$$x_2(t) = \sin^2(2\pi t) \quad ;$$

$$x_3(t) = \delta(t) \quad .$$

   (a) Write down, if possible, a Fourier Series repre-
       sentation for the functions $x_1(t)$, $x_2(t)$,
       $x_3(t)$. If such a representation is impossible
       for any of these functions, explain why.
   (b) Each of the functions $x_1(t)$, $x_2(t)$, $x_3(t)$ is
       applied separately as an input to the linear,
       time-invariant system with frequency response
       function $H(i\omega) = \frac{2}{3+i\omega}$. Find the corresponding
       system responses $y_1(t)$, $y_2(t)$, $y_3(t)$.

63. A "black box" system is known to be linear and time-
    invariant. It is "at rest" prior to application of
    any input. When a unit step function $U(t)$ is ap-
    plied as an input, the output is observed to be

63. (continued)

$$s(t) = tU(t) + (1-t)U(t-1) \quad .$$

(a) Is there enough information to deduce the system
    response to any input? (Explain.) If your an-
    swer is "yes," write down an equation for the
    output $y(t)$ in terms of the input $x(t)$.

(b) Is the system causal? Instantaneous? (Explain.)

(c) Find the impulse response $h(t)$.

(d) Find the system response to input $x(t) = \sin 2\pi t$.
    (Note: This input is $\sin 2\pi t$ for all t,
    $-\infty < t < \infty$.)

64. In the following figure,

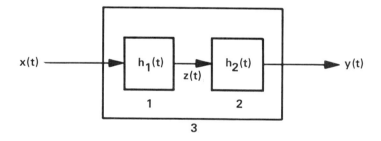

the systems 1 and 2 are linear time-invariant,
and they were completely at rest.

(a) Given $x(t) = U(t)$, $y(t) = (1 + t^2 e^{-t})U(t)$.
    Find $h(t)$, the impulse response function of
    the overall system 3.

(b) Find the differential equation between $y(t)$
    and $x(t)$.

64. (c) If $h_1(t) = e^{-2t}U(t)$, find $h_2(t)$.

65. For the periodic signal shown

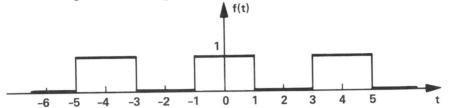

(a) Find the mean square error when $f(t)$ is approximated by

$$\sum_{-2}^{2} F_n e^{in\omega_0 t} .$$

(b) If $f(t)$ is applied to a linear time-invariant
system whose input $x(t)$ and output $y(t)$ are
related by

$$\frac{d^2 y}{dt^2} + y(t) = x(t) , \qquad t > 0$$

$$y(0) = y'(0) = 0 ,$$

find the corresponding output.

66. For each of the following assertions, state <u>true</u> or
<u>false</u>, and also very briefly state <u>why</u> (e.g., briefly
note a definition or property or derivation if true,
or if false give a simple counterexample or a cor-
rection).

(a) The Fourier Transform of an even function of
t is a real and even function of ω.

66. (b) When a signal is convolved with itself, the Fourier Transform of the result is unchanged (equal to transform of the original signal).

(c) The Fourier series of the derivative of a periodic function can be obtained from the series for the function itself by multiplying the nth term by n (for each n).

(d) If, for a linear time-invariant system, the system function or frequency response $H(i\omega)$ is such that $H(0) = 0$, then the output of the system has zero average value.

(e) Suppose a box B, with input signal x and output y, is described by $y(t) = tx(t)$ for all t. Then B is a linear system.

(f) If B_1 and B_2 are two linear boxes, then the two "cascade" connections $C = (B_1$ followed by $B_2)$ and $D = (B_2$ followed by $B_1)$ are equivalent systems (i.e., C and D have the same input/output relation).

(g) The inverse Laplace Transform of a polynomial in s is a linear combination of delta functions (including derivatives of delta functions, as needed).

(h) The impulse response of the system formed by parallel combination of two linear time-invariant systems is the convolution of the two individual impulse responses.

66. (i) The pure sinusoidal input 10 sin 500t, for
         -∞ < t < ∞, is applied to a time-invariant
         black box B. The output is observed to be
         20 sin 100t. We can conclude that the box B
         is not linear.

67. Given 3 systems H_1, H_2, H_3 with impulse re-
     sponses of

$$h_1(t) = U(t) - U(t-1) ,$$

$$h_2(t) = U(t-1) - U(t-2) ,$$

$$h_3(t) = U(t-1) - U(t-3) ,$$

     respectively. Find the overall impulse response h(t)
     of the system:

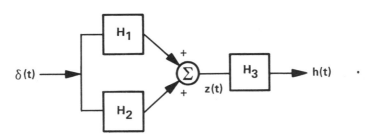

68. Which of the following systems are (i) linear;
     (ii) time-invariant; (iii) zero memory; (iv) causal?

     (a) $y(t) = x(t+3) + x(t-1)$.

     (b) $\dfrac{d^2 y(t)}{dt^2} + y^2(t) = x(t)$.

68. (c) $y(t) = [x(t)][x(t-1)]$.

(d) $y(t) = \int_{0}^{t+1} x(t)\, dt$.

(e) $y(t) = \tan^{-1}\{x(t)\}$.

69. Consider the system shown:

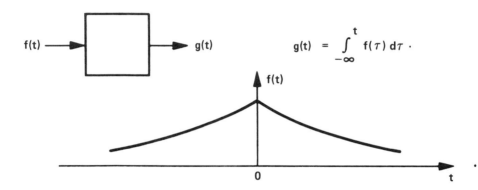

$$g(t) = \int_{-\infty}^{t} f(\tau)\, d\tau \cdot$$

Given that $f(t) = e^{-|t|}$, $-\infty < t < \infty$. Find $G(j\omega)$
-- the Fourier Transform of $g(t)$.

70. Given the first order differential equation

$$\frac{dy(t)}{dt} + y(t) = x(t) \ .$$

Given that $y(0) = 0$ and that

$$x(t) = [U(t) - U(t-4\pi)]\sin t \ .$$

Find $y(t)$ for $t \geq 4\pi$.

71. Let F(iω) be the Fourier Transform of a function

f(t) for −∞ < t < ∞.

(a) When f(t) is an <u>even</u> function of t show
that

$$\frac{dF(i\omega)}{d\omega} = -2 \int_0^\infty t \sin \omega t \, f(t) \, dt \quad .$$

(b) For the function shown:

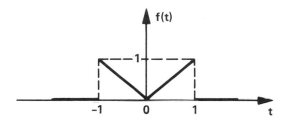

Compute F(iω) and plot the amplitude and phase
spectra of f(t) (i.e., |F(iω)| and θ(ω)).

72. (a) Find f(t) whose Fourier Transform f(iω) is
shown in the figure below

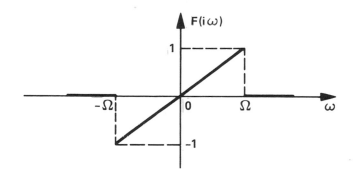

72. (b) If $F(i\omega)$ of (a) is the frequency function of a linear time-invariant system S whose input is $\dfrac{\sin 2\Omega t}{\Omega t}$, show that the corresponding output is proportional to $f(t)$.

73. Consider the system shown

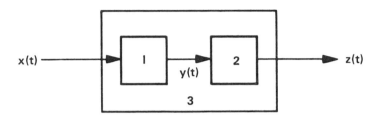

where the system 1 has a system function $H_1(s)$, say, and the system 2 admits the input-output relation

$$z(t) = \int_0^t \sigma y(\sigma)\, d\sigma \ , \qquad t \geq 0 \ .$$

(a) Derive the input-output relation for the system 3 -- that is, the relationship between $z(t)$ and $x(t)$ -- and then show that 3 is time-varying.

(b) If $H_1(s) = \dfrac{s}{s+1}$, compute the impulse response function $H_3(t,\sigma)$ of 3.

74. For the periodic signal shown

74. (continued)

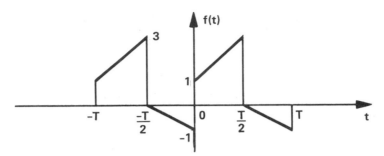

(a) Compute the Fourier coefficients F_n of $f(t)$ and obtain an expression for the amplitude $|F_n|$ and the phase θ_n.

(b) Find the mean square error when $f(t)$ is approximated by sum

$$\sum_{-2}^{2} F_n \, e^{in\omega_0 t} \, ,$$

$T = 4$.

75. Given the system

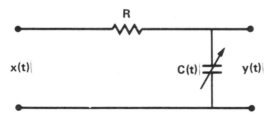

where $x(t)$ and $y(t)$ are input and output voltages as shown. R is a constant and $C(t)$ is a given function of t.

(a) Write down the differential equation relating $x(t)$ and $y(t)$.

75. (b) Solve the equation for y(t) with y(0) = 0 and show that the given system is linear and time-varying.

76. In the following, F(s) is the Laplace Transform of f(t). Given

$$F(s) = \frac{s}{2s^2 + 3s + 1}$$

and

$$f(t) = \int_{-\infty}^{\infty} h(t-\sigma)\ U(t-\sigma)\ e^{-\sigma}\ U(\sigma)\ d\sigma\ ,\quad t \geq 0\ .$$

(a) Find h(t).

(b) Show that f(t) can also be expressed as:

$$f(t) = \frac{1}{2} \int_{0}^{t} e^{-\frac{1}{2}(t-\sigma)}\ [\delta(\sigma) - e^{-\sigma}]\ d\sigma,$$

$$t \geq 0\ .$$

77. For the circuit shown

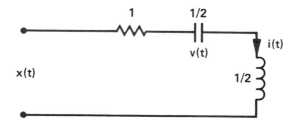

find x(t) -- voltage -- given:

$$i(t) = [\frac{1}{2}\sin 2t - te^{-2t}]\ U(t)$$

152

77. (continued)

   and $v(0) = 0$ -- where v is the voltage across
   the capacitor.

78. Let S be a linear, time-invariant system, at rest
    at $t = 0$. Given the <u>step</u> response

   $$y_U(t) = S[U(t)] = te^{-2t}U(t) .$$

   (a) Find a differential equation relating input
       $x(t)$ to output $y(t)$.
   (b) Find the impulse response of S.
   (c) Find the output of the system for the input
       $x(t) = \sin(4t+\theta)$.

79. In Figure (b) below $y(t)$ is the output of a linear
    time-invariant system when the input $x(t)$ shown in
    Figure (a) is applied to the system.

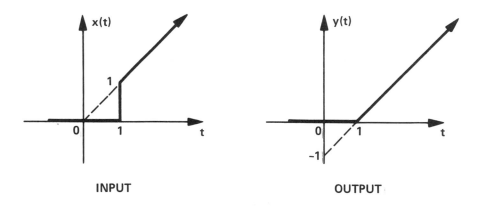

| INPUT | OUTPUT |

   (a) Find the output of the system when the input
       is $\frac{1}{2}(t+1)U(t) + 2(t-1)U(t-2)$.

79. (b) Sketch the output found in (a).

  (c) Find and sketch the first derivative of the
       output in (a).

80. (a) Find the Fourier series expansion and the ampli-
         tude and phase spectra for the periodic function
         f(t) shown:

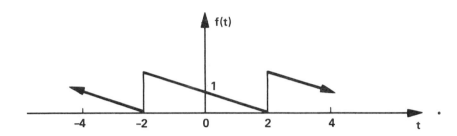

  (b) Find the mean square error when f(t) is
       approximated by:

$$f(t) \cong A_0 + A_2 \cos \omega_0 t + B_1 \sin \omega_0 t .$$

81. (a) Plot the function $f(t) = te^{-t}U(t-1) + \delta(t+1)$.

  (b) This function is now taken to be the input to
       a system whose input-output relation is given
       by

$$y(t) = \int_{-\infty}^{\infty} tU(\sigma-t) \, U(\sigma) \, d\sigma \quad , \quad -\infty < t < \infty .$$

       Find y(t) .

  (c) Show clearly whether the system in (b) is:
       time-invariant or time-varying, causal or not causal.

82. Given

$$e^{-t}, \quad te^{-t}, \quad t^2 e^{-t}, \quad t \in [0, \infty) \quad .$$

(a) Setting

$$\phi_0(t) \;=\; e^{-t}$$

$$\phi_1(t) \;=\; Ae^{-t} + te^{-t}$$

$$\phi_2(t) \;=\; Be^{-t} + Cte^{-t} + t^2 e^{-t} \quad,$$

find A, B and C so that

$$\int_0^\infty \phi_i(t) \; \phi_j(t) \; dt \begin{cases} = 0 & \text{for } i \neq j \\ \neq 0 & \text{for } i = j \end{cases}$$

$$i, \; j = 0, \; 1, \; 2 \quad .$$

(b) Given $u(t) = (1+t)e^{-2t}$, approximate $u(t)$ by

$$u(t) \;\simeq\; \alpha_0 \phi_0(t) + \beta_0 \phi_1(t) \quad .$$

Find α_0 and β_0 and compute the mean square error.

83. Consider the following systems, with input $u(t)$ and output $y(t)$:

(i)

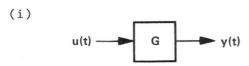

governed by the differential equation

83. (continued)

$$\frac{dy}{dt}(t) + 3y(t) = 2u(t) + 3\frac{du}{dt}(t) \qquad t \in [0,\infty)$$

and with $y(0) = u(0) = 0$.

(ii)

governed by the differential equation

$$\frac{d^2y}{dt^2}(t) + 4\frac{dy}{dt}(t) + 3y(t) = u(t) + K\frac{du}{dt}(t)$$

$$t \in [0,\infty)$$

with $y(0) = \frac{dy}{dt}(0) = u(0) = 0$.

(a) Using the Laplace Transform, obtain expressions for the transfer functions G(s) and P(s), of the above systems.

The two systems described above are now coupled in the feedback form:

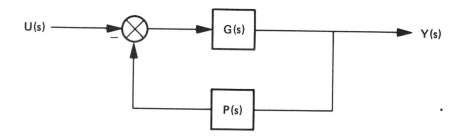

(b) Find an expression for the system function of the feedback system.

(c) For which values of K would this (feedback) system be stable?

84. Find the Fourier Transform of the following functions:

(a) $f(t) = \int_{-2}^{3} e^{-|t-\tau|} d\tau$.

(b) Assuming that the Fourier Transform of $f(t)$ is $F(i\omega)$ find a general expression for

$$F\{f(t) \sin (\omega_k t - \theta)\} \quad , \quad \omega_k, \theta > 0$$

in terms of $F(i\omega)$.

85. The input-output relation of a linear system is

$$y(t) = \int_{-\infty}^{t} 2e^{-(t-1)} \sigma x(\sigma) d\sigma \quad , \quad -\infty < t < \infty,$$

where $x(t)$ and $y(t)$ are input and output, respectively.

(a) Find the impulse response function $h(t,\tau)$ of the system, then show whether the system is time-invariant or time-varying, causal or noncausal.

(b) Find the output of the system when the input $U(t-1)$ is applied.

(c) The output function obtained in part (b) is now applied to a linear time-invariant and causal system -- which was at rest. The following response is observed

$$2(t-1) e^{-(t-1)} U(t-1) \quad .$$

85. (c) (continued)

Find the impulse response function $h(t-\tau)$ of this linear time-invariant and causal system.

86. (a) Find the function $y(t)$ whose Laplace Transform $Y(s)$ is

$$Y(s) = \frac{2s^2}{s^3 + s^2 + s + 1} \; .$$

(b) The function $y(t)$ found in (a) is the output of a linear time-invariant and causal system whose impulse response function is

$$h_1(t-\tau) = \delta(t-\tau) - \sin(t-\tau)U(t-\tau) \; .$$

Find the input $x(t)$ which results in this output $y(t)$.

(c) The linear time-invariant and causal system of part (b) is now cascaded with another linear time-invariant and causal system whose impulse response function is

$$h_2(t-\tau) = \tfrac{1}{2}(t-\tau)^2 U(t-\tau) \; .$$

Show that the output of the cascaded combination to the input $U(t)$ is

$$tU(t) - (\sin t)U(t) \; ,$$

86. (c) (continued)

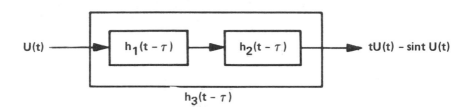

$U(t) \longrightarrow \boxed{h_1(t-\tau)} \longrightarrow \boxed{h_2(t-\tau)} \longrightarrow tU(t) - \sin t\, U(t)$

$h_3(t-\tau)$

Finally show how you would obtain the impulse
response function $h_3(t-\tau)$ of the cascaded combination from the output $tU(t) - (\sin t)U(t)$
above, without using Laplace Transform method.

87. A system S_1 is described by the input–output relation:

$$y(t) = x(t) + 2\frac{\alpha+\beta}{\alpha-\beta}\int_0^t \left[\beta e^{-\beta(t-\sigma)} - \alpha e^{-\alpha(t-\sigma)}\right]$$

$$\times\ x(\sigma)\ d\sigma\ ,$$

$$\text{for}\quad t \geq 0\ ,$$

where α and β are positive constants.

(a) Find the differential equation relating $x(t)$
and $y(t)$. You should consider both cases:
$\alpha \neq \beta$ and $\alpha = \beta$.

(b) The system S_1 is now cascaded with a linear,
time-invariant and causal system S_2 as shown:

87. (b) (continued)

Find the impulse response function of S_2 knowing that the input $U(t)$ results in the output $(2e^{-\alpha t} - 1)U(t)$ -- as shown above.

(c) Find the output of the system S_1 given that $\alpha = \beta$ and $x(t) = tU(t)$.

88. Let $f(t)$ be a periodic function with period $T = 2$, and its amplitude and phase spectra are shown below:

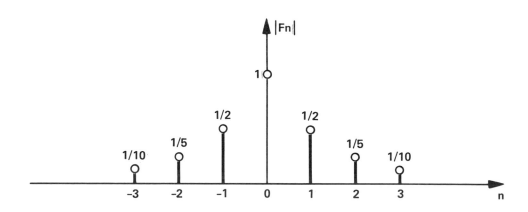

AMPLITUDE SPECTRUM

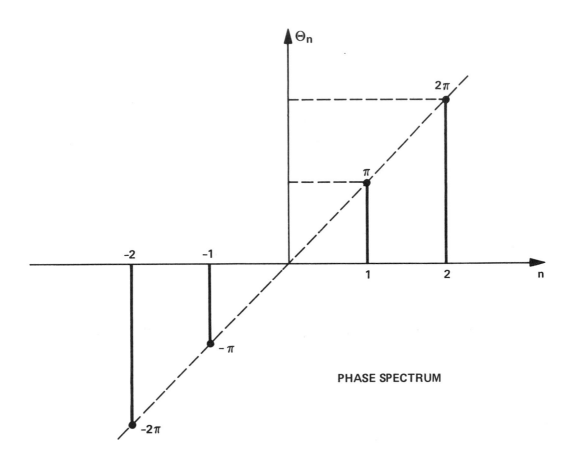

PHASE SPECTRUM

88. (a) Find the Fourier series representation of $f(t)$, and thence $f(t)$.

(b) Find the mean square error when $f(t)$ is approximated by

$$F_1 e^{i\pi t} + F_0 + F_{-1} e^{-i\pi t} .$$

(c) The function $f(t)$ above is now applied to a linear time-invariant system whose impulse response function is

$$h(t) = \frac{\sin \pi t}{2t} - \frac{\sin \left(\frac{\pi}{2}\right)t}{\pi t} , \qquad -\infty < t < \infty .$$

Find the corresponding output.

89. (a) Given the following information:

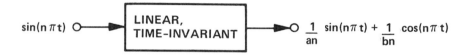

for $-\infty < t < \infty$, $n \neq 0$, and a and b are nonzero constants. Find the output of the same system when the following input is applied to it:

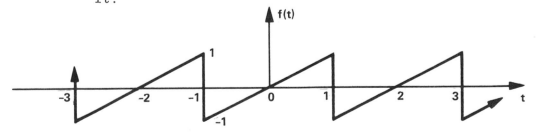

162

89. (b) Find the impulse response function h(t) of a linear time-invariant system S knowing that

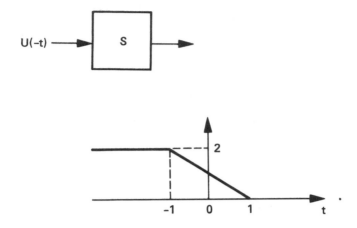

Is S causal? Explain why or why not .

90. Let S be a linear, time-invariant and causal system whose input x(t) and corresponding output y(t) are shown below:

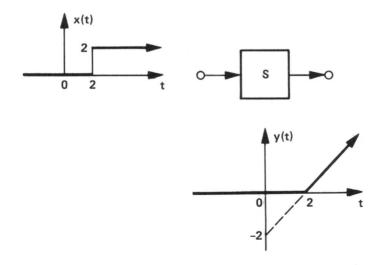

90. (a) Find the impulse response function h(t) of S.

(b) Find the output of S when its input is

$$x(t) = \begin{cases} e^t & \text{for } t < 0 \\ e^{-t} & \text{for } t \geq 0 \end{cases}.$$

91. The periodic signal shown below is the input of a linear time-invariant system whose system function H(s) is

$$H(s) = \frac{2s}{2s+1}$$

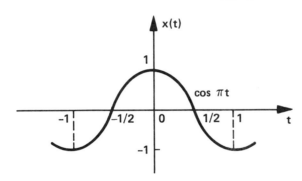

(x(t) over one period).

(a) Find y(t) and $\int_{-1}^{1} |y(t)|^2 \, dt$.

(b) The system of part (a) is cascaded with a second linear system as shown:

91. (b) (continued)

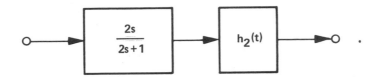

Find the impulse response function $h_3(t)$ of
the combination given that the impulse response
function $h_2(t)$ is

$$h_2(t) = (\tfrac{1}{2} \cos 2t)U(t) .$$

92. Given the differential equation

$$\frac{d^2y(t)}{dt^2} + A\frac{dy(t)}{dt} + By(t) = x(t) , \qquad t > 0 ,$$

where A and B are constants.

(a) Find the constants A and B so that $y(t)$
is expressed in terms of $x(t)$ as

$$y(t) = \int_0^t e^{-(t-\tau)} \sin (t-\tau) x(\tau) d\tau , \qquad t \geq 0,$$

where it is assumed that $y(0) = y'(0) = 0$.

(b) Find $y(t)$ given A = 2, B = 1, $x(t) = \sin t$.
U(t), $y(0) = 0$, and $y'(0) = 2$.

93. For the periodic signal $x(t)$, $-\infty < t < \infty$, shown
on the next page:

93. (continued)

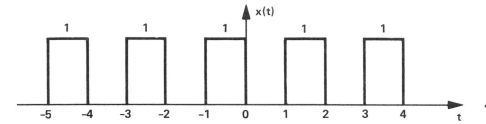

(a) Compute the Fourier coefficients X_n of $x(t)$.

(b) The signal $x(t)$, $-\infty < t < \infty$, is now applied to a linear, time-invariant and causal system whose system function $H(s)$ has the "poles-zeros" plot shown below.

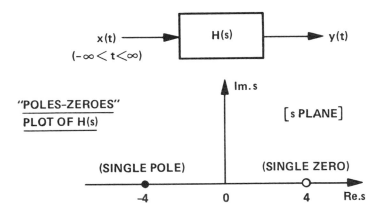

Let $y(t)$, $-\infty < t < \infty$, be the corresponding output of the system, and it is desired to "approximate" $y(t)$ in the mean square sense by the signal:

$$\hat{y}(t) = D_0 + D_1 \sin \pi t + D_2 \cos \pi t, \quad -\infty < t < \infty.$$

93. (b) (continued)

Find D_0, D_1 and D_2 given that $\underline{H(0) = -1}$.
Then compute the mean square error involved in
this approximation.

94. A linear time-invariant system is described by the
differential equation

$$2\frac{d^2y(t)}{dt^2} + 3\frac{dy(t)}{dt} + y(t) = x(t) , \qquad t > 0 ,$$

together with the initial conditions:

$$y(0) = \dot{y}(0) = 0 .$$

(a) Find the impulse response function $h(t-\sigma)$ of
the system.

(b) Find $y(t)$ given that $x(t) = U(t)$, $y(0) = 0$
and $\dot{y}(0) = 1$.

95. The Laplace Transform $F(s)$ of a function $f(t)$ is

$$F(s) = \frac{s-2}{s^2 + 3s + 2} .$$

(a) Show that $f(t)$ can be written as

$$f(t) = e^{-2t}U(t) - \int_0^t 3e^{-(t-\sigma)} e^{-2\sigma} d\sigma , \qquad t \geq 0.$$

(b) If the function $f(t)$ above is <u>the output</u> of
a linear, time-invariant and causal system
corresponding to the input $e^{-t}U(t)$:

167

95. (b) (continued)

$e^{-t} U(t)$ ⟶ [ ] ⟶ $f(t)$,

    find the output of the system when the input is
    $e^{-2t} U(t)$.

96. A linear and causal system S is described by:

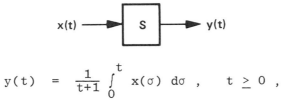

$x(t)$ ⟶ [S] ⟶ $y(t)$

$$y(t) = \frac{1}{t+1} \int_0^t x(\sigma) \, d\sigma \,, \qquad t \geq 0 \,,$$

    where $x(t) = 0$ for $t < 0$.
        Find $y(t)$ given: (i) $x(t) = U(t-1)$;
    (ii) $x(t) = \delta(t-2)$. Is S time-invariant?

97. Find the impulse response function $h(t)$ of a <u>linear</u>,
    <u>time-invariant</u> and <u>causal</u> system S, given:

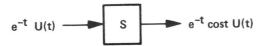

$e^{-t} U(t)$ ⟶ [S] ⟶ $e^{-t} \cos t \, U(t)$

    and

$\delta(t) - e^{-t} U(t)$ ⟶ [S] ⟶ $\delta(t) - e^{-t} (\sin t + \cos t) \, U(t)$.

    <u>Do not use Laplace Transform.</u>

98. Find the system function $H(s)$ and the impulse response function $h(t)$ of a system from the following information:

$$e^{-(t-2)} U(t-2) \longrightarrow \boxed{\begin{array}{l} \text{LINEAR,} \\ \text{TIME-INVARIANT} \\ \text{\& CAUSAL} \end{array}} \longrightarrow \delta(t-2) - (5-2t)e^{-(t-2)} U(t-2)$$

99. (a) Find $f(t)$ given $F(s)$ -- the Laplace Transform of $f(t)$:

$$F(s) = \frac{1}{(s^2+1)^2} \quad .$$

(b) Find the Laplace Transform $Y(s)$ of $y(t)$:

$$y(t) = \int_{-\infty}^{\infty} \tau^2 \sin(t-\tau)\, U(t-\tau) \sin \tau\, U(\tau)\, d\tau \ ,$$

$$t \geq 0 \quad .$$

100. Given:

$$\frac{dy(t)}{dt} + 3y(t) - z(t) = -2x(t) \ , \qquad t > 0$$

$$\text{and} \quad y(0) = 1 \ ,$$

and

$$\frac{dz(t)}{dt} + 2y(t) = -3x(t) \ , \qquad t > 0$$

$$\text{and} \quad z(0) = 0 \ .$$

(a) Denote by $X(s)$, $Y(s)$ and $Z(s)$ the Laplace Transforms of $x(t)$, $y(t)$ and $z(t)$, respectively. Find $Y(s)$ and $Z(s)$ in terms of $X(s)$.

(b) Find $y(t)$ given that $x(t) = U(t)$.

169

BIBLIOGRAPHY

1. R. J. Schwartz and B. Friedland, Linear Systems, McGraw-Hill, New York, 1965.

2. P. M. Chirlian, Signals, Systems, and the Computer, Intext Educational Publishers, New York, 1973.

3. C. D. McGillem and G. R. Cooper, Continuous and Discrete Signal and System Analysis, Holt, Rinehart and Winston, New York, 1974.

4. A. Papoulis, Circuits and Systems: A Modern Approach, Holt, Rinehart and Winston, New York, 1980.

5. A. V. Oppenheim, A. S. Willsky and I. T. Young, Signals and Systems, Prentice Hall, Englewood Cliffs, NJ, 1983.

INDEX